MW00678141

Teach the Way They Learn—Math

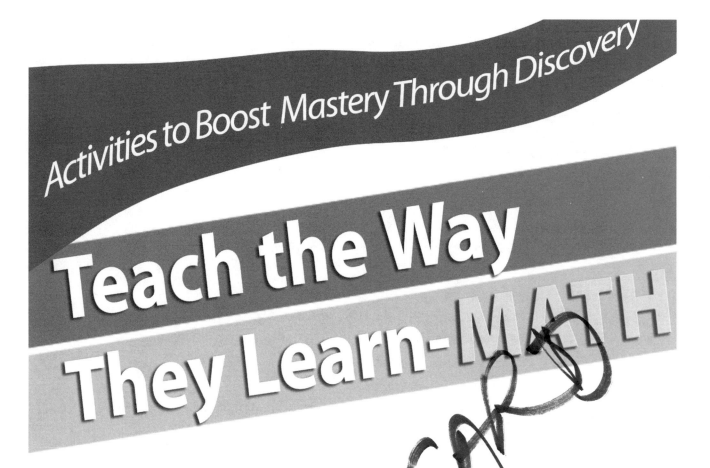

Activities to Boost Mastery Through Discovery

Teach the Way
They Learn—MATH

By Joanne I. Hines and Pamela J. Vincent

Crystal Springs
SDE BOOKS

a division of Staff Development for Educators

Peterborough, New Hampshire

SHELBY COUNTY SCHOOLS
LIBRARY
160 SOUTH HOLLYWOOD
MEMPHIS, TENN. 38112

Published by Crystal Springs Books
A division of Staff Development for Educators (SDE)
10 Sharon Road, PO Box 500
Peterborough, NH 03458
1-800-321-0401
www.SDE.com/crystalsprings

©2008 Joanne I. Hines and Pamela J. Vincent
Illustrations ©2008 Crystal Springs Books

Published 2008
Printed in the United States of America
13 12 11 10   3 4 5 6

ISBN: 978-1-934026-03-8

Library of Congress Cataloging-in-Publication Data

Hines, Joanne I., 1955-

  Teach the way they learn—math : activities to boost mastery through discovery / by Joanne I. Hines and Pamela J. Vincent.

    p. cm.

  Includes bibliographical references and index.

  ISBN 978-1-934026-03-8

  1. Mathematics—Study and teaching (Elementary)—Activity programs. I. Vincent, Pamela J., 1947- II. Title.

QA135.6.H56 2008

372.7--dc22

                              2008011355

Editor: Sandra J. Taylor
Art Director and Designer: Soosen Dunholter
Production Coordinator: Deborah Fredericks
Illustrator: Patrick Vincent

Crystal Springs Books grants teachers the right to photocopy the reproducibles from this book for classroom use. No other part of this book may be reproduced in whole or in part, or stored in a retrieval system, or transmitted in any form or by any means, electronic, mechanical, photocopying, recording, or otherwise, without written permission of the publisher.

Dedication

PB
372.7
Hin
(PROF)
#1300347
Farmington
#17.00

To my wonderful mom, Christine. You always believed in me.
You truly are the wind beneath my wings.
To my incredible sister, Kim. You never cease to amaze me with your inner strength and
sincerity. You make me believe in myself. Thanks for always being there.
To my amazing brother, Rick. You have accomplished so much in so little time.
You make me proud.
Pam Vincent

To my fantastic sister, Helen, and my amazing brother, John. You have both been my
foundation, the strength I need to keep going, and the love that I need to soar in life.
The three of us will always be one.

Joanne Hines

Acknowledgments

We would like to acknowledge the following people:
Thanks to Lorraine Walker and Crystal Springs Books
for making our dreams come true again.
Thanks to Sandy Taylor, our fantastic editor, for her enduring patience
and her great suggestions.
Thanks to Patrick Vincent, who worked his magic with the illustrations for the third time.
Thanks to the outstanding teachers we work with—
for your friendship, support, and expertise.
And finally, thanks to our delightful students. You have taught us so much!

Contents

Skills Addressed and Suggested Grade Levels

TITLE	SKILL	K	1ST	2ND	3RD	4TH	5TH	6TH
Bugs in a Bag	Addition			X	X	X	X	X
Doing the Diamond	Addition/mental math		X	X	X	X	X	X
Fast Fingers	Addition/number sense		X	X	X	X	X	X
Pip, Pip Hooray!	Addition/mental math		X	X	X	X	X	X
Scrambled Eggs	Addition		X	X	X	X	X	X
Tower of Pips	Addition		X	X	X	X	X	X
The Trouble with Doubles	Addition			X	X	X	X	X
Seeing Spots	Addition		X	X	X	X		
Brain Drain	Addition/subtraction			X	X	X	X	X
Toss, Tumble, and Solve	Addition/subtraction		X	X	X	X	X	X
Zip It!	Addition/subtraction	X	X					
The Fun Fan for Math	Any math skill			X	X	X	X	X
Egg-straordinary…Activities	Counting/number sense		X	X				
How Many Toes Do You See?	Counting/number sense	X	X	X				
Pocket Pal Match-Ups	Counting/coin recognition	X	X	X	X	X		
The Great Glyph	Deductive reasoning			X	X	X	X	X
One Hundred Hungry Ants	Division			X	X	X		
Estimate, Mate!	Estimation/counting/multiplication				X	X	X	X
Expanded Notes	Expanded notation				X	X	X	X
We Are Family	Fact families		X	X	X			
Factors and Multiples	Factors/multiples					X	X	X
Let's Compare Fractions	Fractions		X	X	X	X	X	X
Fractions are Fun	Fractions			X	X	X		
Licorice Stick Geometry	Geometry		X	X	X	X	X	X
Geometry Field Trip	Geometry				X	X	X	X
All Sorts of Things	Graphing/categoriziing		X	X	X	X	X	X
The Talking Graph	Graphing		X	X	X	X		
Let's Make a Deal!	Greater than/less than	X	X	X	X	X		
How Big Is a Foot?	Measurement			X	X	X		
Take Away or Tally Up	Mental math			X	X	X	X	X
Cups o' Beans!	Multiplication				X	X	X	
Multiplication Mania	Multiplication			X	X	X	X	

Skills Addressed and Suggested Grade Levels (continued)

TITLE	SKILL	K	1ST	2ND	3RD	4TH	5TH	6TH
Multiplication Circles	Multiplication			X	X	X		
The Magic Flasher	Multiplication/division		X	X	X	X	X	X
Bundles	Multiplication/division				X	X	X	X
You Choose, I Choose	Number order/greater-less than		X	X				
Toss On, Toss Off	Number sense		X	X				
Happy Trails Hundred Chart	Number sense		X	X	X	X		
Math About Me	Number sense		X	X	X	X	X	
Super-Special Action Subtraction	Number sense/subtraction	X	X	X	X	X	X	X
Numbers on the Move	Number sense	X	X	X	X	X		
Fill in the Blanks	Number sense	X	X	X	X	X	X	X
Let Me See, Which Number . . .	Number sense/numerical order	X	X	X	X			
Hundred Chart Puzzles	Number sense/problem -solving		X	X	X			
Race to 100	Numerical order	X	X	X	X	X		
Bears in Line	Ordinals	X	X	X				
Show Me Your Numbers	Place value/fact families		X	X	X	X		
The Sock Sale	Problem-solving				X	X	X	X
Puzzled	Problem-solving	X	X	X	X	X	X	X
Problem of the Day	Problem-solving		X	X	X	X	X	X
Hundred Chart Magic	Problem-solving		X	X	X	X	X	X
Skittles Ratios	Ratios/fractions		X	X	X	X	X	X
Cube It!	Reinforcement of math skills		X	X	X	X	X	X
Math Scavenger Hunt	Reinforcement of math skills		X	X	X	X	X	X
Extra, Extra! Read All About It!	Reinforcement of math skills					X	X	X
Have a Ball	Reinforcement of math skills	X	X	X	X			
Aim for the Stars	Subtraction	X	X	X				
Sing Me a Subtraction Rhyme	Subtraction			X	X	X	X	X
The Human Clock	Time			X	X	X	X	
Time-Telling Teams	Time	X	X	X	X			

Introduction

The math curriculum in schools today has undergone some major changes. It now includes more mental computation and concept mastery through hands-on learning. In addition, students are expected to incorporate written responses for comprehension and assessment purposes. In order to write proficiently and accurately, students must possess a math vocabulary and have practice writing about math. The National Council of Teachers of Mathematics (2000) has also proposed that two of our chief goals as teachers should be to convince students that they can do math and to help them enjoy it.

These challenges inspired us to write *Teach the Way They Learn—Math*. Focusing more on mastery through discovery rather than simply memorization of facts, our book is filled with fun yet meaningful activities. Each one uses interactive groups or pairs in a way that includes social interaction, less frustration, and greater success for the students. The motivating activities foster discussion, develop reasoning skills as well as good listening habits, and actively engage students in learning mathematics.

Clear illustrations and time-saving reproducibles are found on every page of our book, and a skills chart in the front provides you with an easy-to-use reference source. Glance through it to find out which activities provide practice with which math skills and for what age group.

Accompanying the activities are suggested vocabulary words. Since word walls are very popular in classrooms—but are used mainly for reading—we decided to emphasize math vocabulary in a similar way. The difference is that students have their own portable math word wall. You simply supply each student with a manila folder and the four reproducible pages in the back of this book (see pages 104–107). Students glue each page, in alphabetical order, to the four sides of their manila folder, creating their own personal Math Word Folder to keep in their desks and refer to as needed.

Every student is required to make decisions, use manipulatives, become familiar with mathematical language, and respond to Journal Entry questions or statements. The latter will help them build writing skills and encourage critical thinking and problem-solving.

It's time for students to look forward to math class, and with this new resource they will—and so will you!

Bugs in a Bag

Two-digit addition is "in the bag" with this easy and entertaining partner activity.

MATERIALS

- Bugs reproducible (see next page)
- 3 resealable plastic bags per pair of students

PREPARATION

Duplicate the Bugs reproducible and write a single-digit number (0 to 9) on them. Cut them out and divide them equally among the three bags.

STEP BY STEP

Group students into pairs and distribute the materials. Have both students in each pair draw, in turn, one bug from each of the three bags. If duplicates are drawn, the student returns the duplicate bug to the bag and draws another bug. Then one of the pair uses her three numbers to make as many two-digit-plus-one-digit addition equations as she can. Working together, the paired students solve a total of 12 problems.

For example, if one student has the numbers 7, 5, and 3, she might write: 75 + 3, 57 + 3, 53 + 7, 35 + 7, 37 + 5, 73 + 5. Once the partners have each written their problems, they do the addition to complete the equation.

VARIATION/EXTENSION

Have the students do subtraction problems based on the numbers they have in their bags.

JOURNAL ENTRY

Create an addition word problem using your three numbers.

GRADES: *2–6*   **SKILL ADDRESSED:** *Addition*   **VOCABULARY:** *Digit*

ESSENTIAL QUESTION: *How many problems can I make using three numbers?*

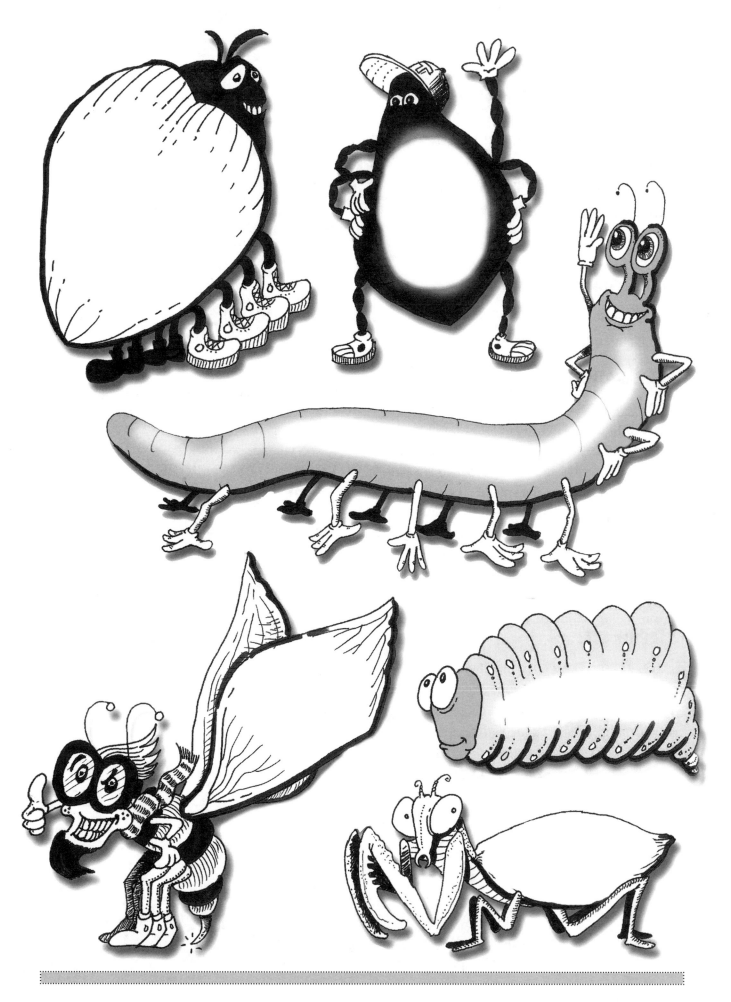

Doing the Diamond

Here's a great way to practice mental math and make better use of your students' time when they are standing in line. It also works as a ticket-out-the-door activity.

MATERIALS

- 22 x 28-inch sheet of tagboard
- Nonpermanent marker

PREPARATION

Draw a diamond pattern (see illustration) on the tagboard and laminate before writing on it. Write a number at each corner of the diamond, another number inside the square, and another on one of the lines of the square. There should be six numbers in all. Use the numbers 1 to 6 first, 2 to 7 next, and then 4 to 9 as the students' mental math prowess increases. Post your diamond near the classroom entrance in easy view of the students when they are lining up.

STEP BY STEP

Model and practice the directions for problem-solving by "doing the diamond." For example, you might say: "Follow my directions for 'doing the diamond.' First, find the sum of up and down [students add the top number, 4, and the bottom number, 7]. Next, find in and left [students add the number inside the square, 3, and the number on the left, 5]. Then find on and right [students add the number on the line of the square, 2, and the number on the right, 6]."

This practice activity for mental math also promotes skills such as directionality, listening, and sequencing.

VARIATIONS/EXTENSIONS

Use multiplication instead of addition.

Use three directions for multiple addends.

JOURNAL ENTRY

Choose a mystery object in the classroom. Describe it but do not name it. Write the directions to find it. Trade with a buddy and see if he can follow your directions to the mystery object and if you can follow his.

GRADES: *1–6*   **SKILLS ADDRESSED:** *Addition, mental math*   **VOCABULARY:** *Diamond, mental math, directionality, left, right*

ESSENTIAL QUESTION: *How can I use the diamond to help me with mental math?*

Fast Fingers

This 5-minute, 10-finger activity is a quick yet effective way to help students develop number sense and master addition facts.

MATERIALS

No materials necessary

PREPARATION

None required

STEP BY STEP

Divide students into pairs and number them as ones and twos. Tell partners to put both hands behind their backs and that when you say "Fast Fingers," they move one hand to the front with fingers showing a number. Tell the Number Ones to call out the greater number of the two hands shown. Repeat and have Number Twos identify the greater number.

During another class time, use the same activity but have students identify the lesser number. For a more advanced version, have partners show fingers on both hands and identify greater and lesser.

As students become more proficient, ask them to add the fingers shown on one hand of both partners and take turns calling out the answer. Finally, have students add the fingers on both hands and take turns calling out the answers.

VARIATIONS/EXTENSIONS

Ask first-grade students to simply recognize the number of fingers shown by their partner.

Ask third- and fourth-grade students to multiply the two sets of fingers.

JOURNAL ENTRY

Draw the outline of your hand. On each finger, write one fact about you that includes a number, such as, "I am 7 years old. I have 2 cats."

GRADES: *1–6*   **SKILLS ADDRESSED:** *Addition, number sense*   **VOCABULARY:** *Numbered heads*

ESSENTIAL QUESTION: *How can I use my fingers to help me add two numbers without counting?*

Pip, Pip, Hooray!

Differentiate your instruction with this ongoing activity that increases in difficulty as your students advance in mental math mastery.

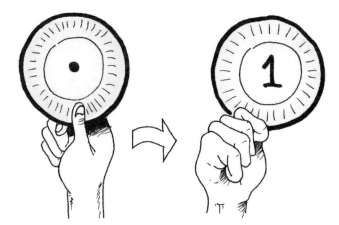

MATERIALS (FOR EACH STUDENT)

- 12 thin, white, inexpensive dessert-size paper plates
- 1 bingo dauber

PREPARATION

None required

STEP BY STEP

Distribute the materials. Student preparation of the plates is a valuable part of the lesson, so model what to do as you give directions to the students. Using a bingo dauber, mark one pip on a paper plate and hold up your plate to show students where to mark their pip (see illustration). Write the numeral 1 on the back of the plate. On another paper plate, mark two pips and hold up your plate to show students the configuration to use. Write the numeral 2 on the back of the plate. Continue the same procedure through numeral 6. Then create a second set of six plates with the same configuration for the pips. Do the plate preparation as one day's math lesson.

The next day, have students spread out one set of plates in front of them, pip side up, and in order from 1 to 6. Explain to them that when you call out a number, they will hold up the plate with that number of pips on it. Continue until all students are familiar with the configurations without having to count them. The next day, add the second set of plates. Hold up a plate with any numeral on it and the plate with numeral 1. For example, hold up numerals 5 and 1. Model how to add the numbers together: 5 + 1 = 6. Tell your students to hold up the same two plates, add the two numbers together, and say the answer. Continue the same procedure, using any numeral and the numeral 2 plate. Ask students to hold up the same plates and say the answer.

GRADES: *1–6*   **SKILLS ADDRESSED:** *Addition, mental math*   **VOCABULARY:** *Counting on*

ESSENTIAL QUESTION: *How can I use domino pips to help me master addition facts 1 to 12?*

Move on to practicing double facts (2 + 2, 3 + 3, etc.). This time hold up two plates with the same number of pips and ask your students to hold up two plates with the same number. As students progress, name a sum and have them hold up two plates that make that sum. Since there will be more than one correct answer, students should record their combinations on a piece of paper.

Now students should be ready for counting on. Hold up one numeral plate and one plate showing pips. Ask students to use both the number sides and pip sides of their plates and hold up the same plates you are holding. Then ask students to add the two numbers on their plates and say the answer.

Hold up two plates showing numbers only. Have students hold up the same two plates, add them together, and say the answer. Next, hold up one number and ask students to find two plates, number sides only, that equal that sum. Since there will be more than one combination, record the problems on the chalkboard or overhead transparency as students share their two-number combinations. For example, if you hold up 6, students may hold up 1 and 5, 3 and 3, or 4 and 2.

VARIATION/EXTENSION

Use this differentiated activity with a small group of children rather than the whole group.

JOURNAL ENTRY

Draw two plates. Write the number 6 on one plate and make four pips on the other. Explain how you would find the sum of these two plates.

Scrambled Eggs

This egg-carton activity helps students learn how to formulate and write addition facts.

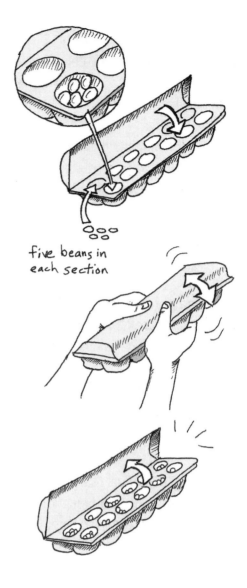

five beans in each section

MATERIALS (FOR EACH PAIR OF STUDENTS)

- 1 egg carton
- 60 dried kidney beans (or any other dried bean that's easy to manipulate)
- Paper and pencils

PREPARATION

- Assemble the materials.

STEP BY STEP

Divide students into pairs. Give each pair an egg carton and a handful of dried beans. Have students count out 60 beans and place 5 in each egg-carton cup. One partner closes the lid securely and shakes the carton several times. When the carton is opened, students count the beans in the two cups across from each other and write an equation on their papers using the two numbers. If one cup is empty, then that number would be zero. There are six possible equations, but point out that students should not write duplicate equations.

In round two, the other partner shakes the carton and both repeat the procedure. Students work together to write the equations. You may want to make a master chart of all equations formed.

VARIATION/EXTENSION

Give the class a number in advance. Have the partners subtract the number of beans in each egg-carton cup from that number. There are 12 possible equations.

JOURNAL ENTRY

Write two examples of how you would use addition in real life.

GRADES: *1–6*   **SKILL ADDRESSED:** *Addition*   **VOCABULARY:** *Total, equation, section*

ESSENTIAL QUESTION: *How can I use beans to make an addition equation?*

Tower of Pips

Students sort and line up dominoes according to their sums and, at the same time, create a cityscape.

MATERIALS (FOR EACH PAIR OF STUDENTS)

- Set of dominoes
- Sentence strip, 24 inches long

PREPARATION

Write the numbers 0 through 12 on each sentence strip, spaced evenly along its length.

STEP BY STEP

Divide students into pairs. Give each pair a set of dominoes and a numbered strip. All dominoes should be face up. Tell students to take turns choosing a domino, adding the pips, and placing it above the correct number on their strip (see illustration). Dominoes are laid flat on the floor or table and placed end to end above the numbers on the strip. Upon completion, the dominoes will resemble towers or skyscrapers in a city rising up from a street, represented by the strip. The white pips will look like lights in the buildings. After the towers are in place, discuss which ones had the most and least dominoes.

VARIATION/EXTENSION

Use this information to create a bar graph.

JOURNAL ENTRY

Which tower had only 1 domino? Which tower had the most dominoes? Why?

GRADES: *1–6*   SKILL ADDRESSED: *Addition*   VOCABULARY: *Pips*

ESSENTIAL QUESTION: *Which sum tower will have the greatest number of dominoes?*

The Trouble with Doubles

This double-trouble game reinforces the strategy of "doubles + 1." Students can use what they know (doubles facts) to determine what they don't know.

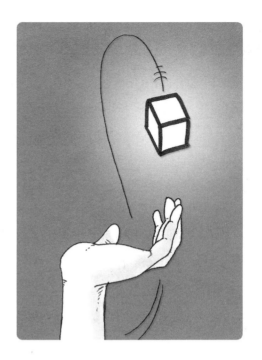

MATERIALS

- Blank wooden cube for each pair of students
- Trouble with Doubles game board reproducible (see next page) for each pair of students
- 4 counters of a different color for each student

PREPARATION

Write one of the following numerals on each side of the cubes: 3, 4, 5, 6, 7, or 8. Duplicate the game board and gather the colored counters.

STEP BY STEP

Distribute the materials. Explain to the students that the goal of the game is to be the first to get all four of his counters on the game board. Divide the class into pairs. The first player rolls the number cube, doubles the number rolled, and adds 1 to that number. For example, if Mark rolled an 8, he doubles the number (8 + 8 = 16) and then adds 1 (16 + 1 = 17). If the number 17 is on his game board, he covers the answer with a counter and rolls the cube again. If the answer is not on his board, the next player takes a turn.

VARIATION/EXTENSION

Students write the equation on paper after they place the counter on the answer and explain their strategy. For example, if Lilly covers 15, she would write 7 + 8 = 15 and explain that 7 + 7 = 14 and 14 + 1 = 15.

JOURNAL ENTRY

I can think of many things that come in doubles.

GRADES: *2–6*   **SKILL ADDRESSED:** *Addition*   **VOCABULARY:** *Doubles, plus*

ESSENTIAL QUESTION: *How can I use doubles facts plus 1 to help me determine related facts?*

Seeing Spots

Partners practice addition facts in this game of dots and spots.

MATERIALS (FOR EACH PAIR OF STUDENTS)

- Dalmatian reproducible (see next page)
- Set of dominoes
- Plastic chips or markers

PREPARATION

Duplicate the reproducible, making enlarged copies for younger students, if desired.

STEP BY STEP

Divide students into pairs. Give each pair a copy of the reproducible, a set of dominoes, and some chips or markers. Ask them to turn their dominoes face down. Students take turns turning over a domino, adding the dots shown, and covering the spot on the Dalmatian that has the appropriate sum with a plastic chip. Play continues until all spots are covered.

VARIATION/EXTENSION

Use numbered cubes to toss instead of dominoes for students who are more proficient with addition facts.

JOURNAL ENTRY

Continue this pattern:
2, 3, 5, 8, 12, ___, ___, ___, ___.
Explain your strategy.

GRADES: *1–4*   **SKILL ADDRESSED:** *Addition*   **VOCABULARY:** *Face down*

ESSENTIAL QUESTION: *How can I use dominoes to help me learn addition facts?*

Brain Drain

Help students make a mental connection between addition and subtraction while playing this pre-algebra game.

MATERIALS (FOR EACH GROUP OF THREE STUDENTS)

- 2 copies of the Brain Cards reproducible (see next page)
- Paper of two different colors

PREPARATION

Copy the reproducible on two different colors of paper. Laminate and cut out each card.

STEP BY STEP

Divide the class into groups of three. Separately shuffle each set of cards (or let students shuffle them) and place them in two piles, face down, between two players who are sitting across from each other. Tell the two students to simultaneously pick up a card from the top of their pile and, without looking at their cards, hold them face outward on their foreheads. The third player, who can see both numbers, adds them together and tells the other two players the sum of the two cards. Those players use the sum as well as the numeral on their partner's forehead (which they can see) to help them determine the numeral on their forehead (which they cannot see).

For example, Sue shows 5 on her forehead and Ted shows 2. Toni tells them the total is 7. Sue says her card is 5 because she knows 7 − 2 = 5. Ted says his card is 2 because 7 − 5 = 2. When all cards have been used, they are reshuffled and placed again in separate stacks. Players change roles and continue. There is no winner and no competition in this game.

JOURNAL ENTRY

Sometimes math can be a headache. Explain.

GRADES: *2–6*   SKILLS ADDRESSED: *Addition, subtraction*   VOCABULARY: *Whole, part, mental math*

ESSENTIAL QUESTION: *How can I figure out the sum if I only know one part?*

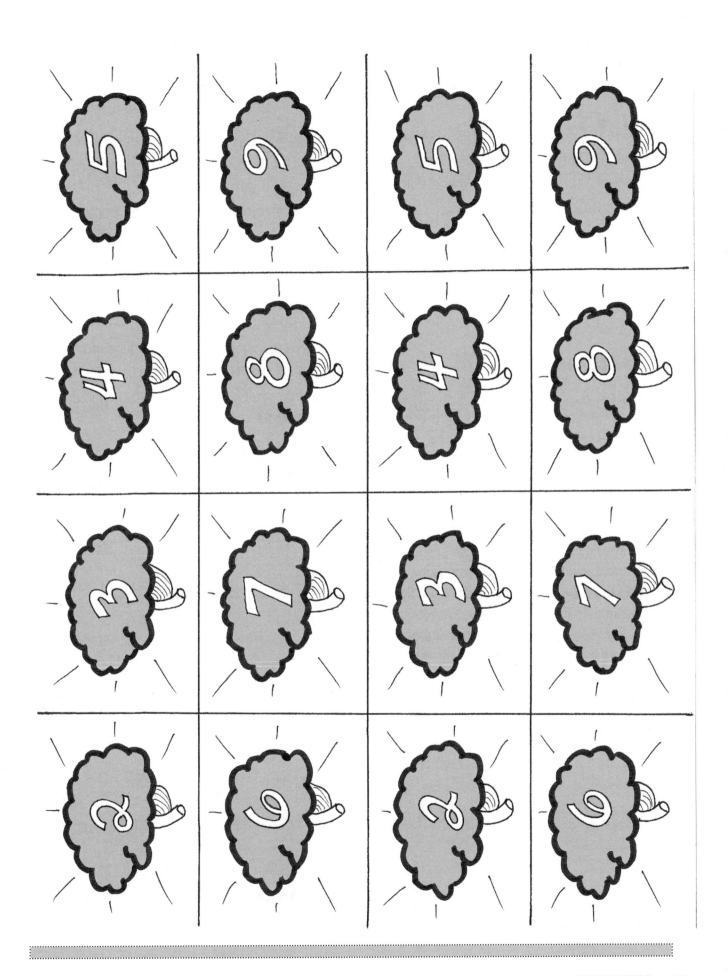

Toss, Tumble, and Solve

Solving addition and subtraction equations is the name of the game in this winning activity.

MATERIALS (FOR EACH PAIR OF STUDENTS)

- 3 blank, 1-inch wooden cubes
- Paper cups

PREPARATION

Prepare the cubes by writing 1, 2, 3, 4, 5, and 6 on two of them and three plus signs (+) and three minus signs (–) on the third one.

STEP BY STEP

Divide the students into pairs. Give each pair a paper cup, two numbered cubes, and one operations cube. Ask them to take turns placing their cubes inside the paper cup, shaking it, and tossing the cubes onto a tabletop or the floor. Then students take turns solving the equation shown. If the minus sign is thrown, students subtract the smaller number from the larger. If the plus sign is thrown, students add the two numbers. Students continue until both partners have had 10 turns. Decide whether the students should write each equation or just keep a running tally of the number of equations each one solves correctly.

VARIATIONS/EXTENSIONS

Use one die in place of one numbered cube. This may help students who have not yet mastered the addition facts to understand the concept of counting on.

Add the multiplication sign to the operations cube for older students.

JOURNAL ENTRY

Write one addition and one subtraction word problem. Illustrate both.

GRADES: *1–6*   **SKILLS ADDRESSED:** *Addition, subtraction*   **VOCABULARY:** *Addition, subtraction*

ESSENTIAL QUESTION: *How can I practice addition and subtraction facts with one activity?*

Zip It!

Zippers are a nifty and novel tool for helping students master addition and subtraction.

MATERIALS

- 22-inch-long coat zippers, one for each pair of students
- Paper and pencil for each student

PREPARATION

Create a working number line on each zipper by writing the numbers 0 to 10 along its length (see illustration).

STEP BY STEP

Divide the class into pairs and distribute the materials. Ask the students to find the sum of 4 + 3, using the zipper like a number line. Students unzip to the 4 and then zip up 3 places, ending at the 7. Students record the problem and answer on their paper. Continue with four more examples, showing addition as zipping up. Students work with their partners to solve problems together.

During another class period, ask the students to work on the problem of 10 − 7. Students place the zipper on 10 and zip down 7 places, ending at 3. Students record the problem and answer on their paper. After several examples, they work with their partners to solve subtraction problems together.

Continue to supply zippers to students who need a manipulative to solve problems.

VARIATIONS/EXTENSIONS

Let kindergartners use the zipper to identify numbers greater than and less than.

Make longer number lines from 0 to 20.

JOURNAL ENTRY

Pretend you could walk on the number line. How many steps would you take if your problem was 4 + 5? On which number would your walk end?

GRADES: *K–1*   **SKILLS ADDRESSED:** *Addition, subtraction*   **VOCABULARY:** *Up, down*

ESSENTIAL QUESTION: *How can I use a zipper number line to help me understand the processes of addition and subtraction?*

The Fun Fan for Math

This Fun Fan is the math version of the one in our book *Teach the Way They Learn* for language arts activities. Once made, it can be used over and over again.

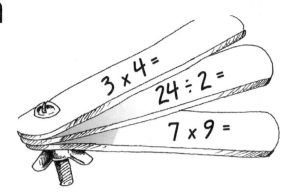

MATERIALS

- 20 large craft sticks for each student
- Drill with 1½- to 2-inch diameter bit
- Small, resealable plastic bag for each student
- Permanent marker
- 1 wing nut for each student
- 1 bolt (1½- to 2-inch diameter) for each student
- 1-inch square gum eraser for each student
- #2 pencil for each student

PREPARATION

In the end of each craft stick, drill a hole the same diameter as the bolt. Write the name of a student in permanent marker on the outside of the plastic bag and place 20 craft sticks inside the bag, along with a wing nut, bolt, pencil, and eraser. Repeat the process for each student.

STEP BY STEP

Distribute the materials. Have students remove 10 of the 20 sticks in their bags. Tell them to write, in pencil, their most difficult multiplication, division, addition, and/or subtraction math facts on the front of each stick and the answer to each on the back of the stick.

Ask them to attach the sticks with the nut and bolt to create their Fun Fan (see illustration). Then have them exchange fans with their neighbors and use them like flash cards.

When the activity is completed, have each student disassemble his fan, erase the writing on both sides of each stick, and return all the sticks and the nut and bolt to the plastic bag. Materials will now be ready for the next time you want to do the activity.

VARIATION/EXTENSION

Use these fans for assessment: Place 10 mini clocks, numbered 1 to 10, on the chalkboard ledge. Ask students to write on the first stick the time shown on the first clock; on the second stick, the time on the second clock; and so forth. Then have them attach the sticks with the nut and bolt, place their fan in the bag, and put it in a container for you to check. You can do this easily by fanning out the sticks.

JOURNAL ENTRY

What is the hardest thing for you to do in math? Explain.

GRADES: *2–6*   **SKILL ADDRESSED:** *Any math skill*   **VOCABULARY:** *Varies based on skill*

ESSENTIAL QUESTION: *How can I practice my math skills and monitor my progress?*

Egg-straordinary, Egg-ceptional, and Egg-cellent Activities

Teach students about number words, counting, and number sense and recycle your egg cartons at the same time!

MATERIALS (FOR EACH PAIR OF STUDENTS)

- Egg carton
- 78 beans or markers

PREPARATION

Write inside each individual egg-carton cup the words *one* through *twelve* in order.

STEP BY STEP

Divide the students into pairs and distribute the materials. Ask students to take turns counting and placing the specified number of beans in each egg-carton cup based on the number word written inside. If counting is accurate, the students will use "eggs-actly" 78 beans.

VARIATION/EXTENSION

Write the number names inside the egg carton cups in random order.

JOURNAL ENTRY

Eggs are sold in a dozen. What are some other things that are sold or come packaged in a dozen?

GRADES: *1–2*   **SKILLS ADDRESSED:** *Counting, number sense*

VOCABULARY: *One, two, three, four, five, six, seven, eight, nine, ten, eleven, twelve, dozen*

ESSENTIAL QUESTION: *What are one, two, three...?*

How Many Toes Do You See?

Kids count toes, instead of their fingers, as they practice counting by fives. Fun for your visual learners too!

MATERIALS

- Foot pattern reproducible (see next page) for each student

PREPARATION

Duplicate the foot pattern, which includes the following:

> Wiggly, wiggly toes I see,
>
> On giggly, giggly friends and me.
>
> Count by five and you will see,
>
> How many toes that there can be!

STEP BY STEP

Give each student a copy of the foot pattern. Ask students to draw a picture of themselves and several friends in their bare feet at the top of the pattern. Then have them complete the sentence at the bottom of the foot, which reads, "There are _____ toes in all!"

VARIATIONS/EXTENSIONS

For added fun, allow the students to paint the toenails with real nail polish! (Ask your favorite "nail tech" to share her old polish.)

Display the footprints on the wall in a trail around the classroom.

Put all student papers into a class foot book.

JOURNAL ENTRY

I can count by 5s to 100.

GRADES: *K–2*   SKILL ADDRESSED: *Counting*   VOCABULARY: *Fifteen, twenty, twenty-five, thirty, thirty-five*

ESSENTIAL QUESTION: *How can I use my toes to count by fives?*

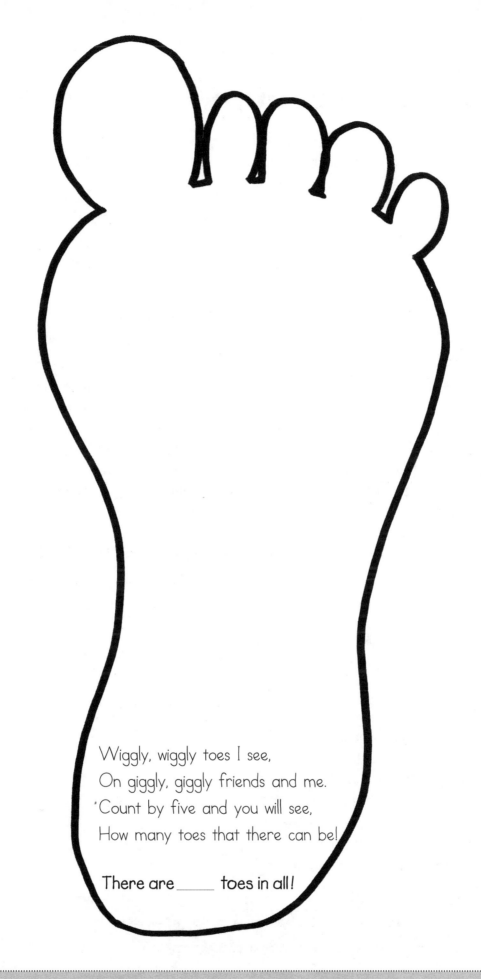

Wiggly, wiggly toes I see,
On giggly, giggly friends and me.
Count by five and you will see,
How many toes that there can be!

There are _____ toes in all!

Pocket Pal Match-Ups

Everybody likes to have a little change in their pockets. This activity requires kids to count that change and recognize equal coin combinations.

MATERIALS (FOR ALL STUDENTS)

- Pocket Pattern reproducible (see next page)
- 8½ x 11-inch sheets of construction paper
- Plastic coins

PREPARATION

Duplicate the reproducible. Cut the construction paper in half crosswise, creating two 8½ x 5 ½-inch sheets. Make enough so each student has a half sheet.

STEP BY STEP

Give each student a sheet of the halved construction paper and a copy of the pattern. Model how to make the pocket and then ask the students to construct their own. Once they have completed their pockets, have them place theirs in a designated box or area at the end of the day.

Decide what amounts of money the students will be counting. Place combinations of coins that equal that amount in groups of four pockets. For example, using the amount of 25 cents, one group of four pockets might include 1 quarter; 2 dimes and 1 nickel; 25 pennies; and 5 nickels. Using 50 cents, another group might include 2 quarters; 5 dimes; 1 quarter, 2 dimes, and 1 nickel; and 10 nickels. Fill the pockets before the beginning of class the next day.

The next day have each student count the coins in her pocket and then find the other students who have the same amount in their pockets. Once together, the students in the group count their coins for each other and present their combinations to the entire class. Collect the pockets and store for use at another time.

VARIATIONS/EXTENSIONS

For kindergartners, who are learning coin recognition as well as value, place the same coin in four pockets.

For grades 3 to 4, use higher amounts that include paper bills of $5, $10, and $20.

JOURNAL ENTRY

This is how I spent my $5 allowance….

GRADES: *K–4*   **SKILLS ADDRESSED:** *Counting, coin recognition*   **VOCABULARY:** *Change*

ESSENTIAL QUESTION: *What different coins can I use to show the same amounts of money?*

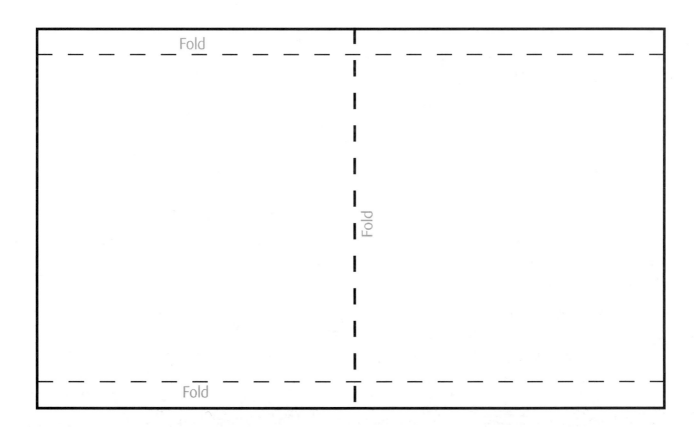

The Great Glyph

A glyph—short for hieroglyphics—is a form of picture writing that conveys information. In this activity, students create a visual that makes a wordless mathematical statement about personal facts and opinions.

MATERIALS (FOR EACH STUDENT)

- 8½ x 11-inch sheets of construction paper in red, blue, green, and yellow
- Markers
- Scissors
- Glue

PREPARATION

Cut the sheets of construction paper in half so that you have four 5½ x 8½-inch sheets, one of each color, for each student. Write on the board the criteria the students will be using to make their glyphs. This could be as follows:

Body Shape: If you are a boy, cut out a triangle from the blue construction paper. If you are a girl, cut out a circle from the green construction paper.

Eyes: If you are wearing jeans, cut out 2 oval eyes from the yellow construction paper and draw a dot in the bottom of each. If you are not wearing jeans, cut out 2 oval eyes and draw the dot in the top.

Legs and Shoes: If you like math, make your arms, legs, and shoes green. If you do not like math, make your arms, legs, and shoes red.

Arms: If you have light hair, make your arms reaching up. If you have dark hair, make your arms hanging down.

STEP BY STEP

Distribute the materials. Explain to the students that they will be using the construction paper, markers, scissors, and glue to create a character that gives information about themselves. Instruct the students to use the criteria you have written on the board.

Have students assemble their bodies by gluing the various parts together (see illustration), adding other features, such as a nose and mouth, and decorating their clothes. Then have the class show their glyphs and discuss what they tell about that student.

VARIATION/EXTENSION

Use different characteristics and features for the Great Glyph key (next page) and give a copy to each student.

JOURNAL ENTRY

Write about the information your glyph tells.

GRADES: *2–6*　　　**SKILL ADDRESSED:** *Deductive reasoning*　　　**VOCABULARY:** *Glyph*

ESSENTIAL QUESTION: *How can you use a picture recipe to tell about yourself?*

The Secret KEY to the Great Glyph!

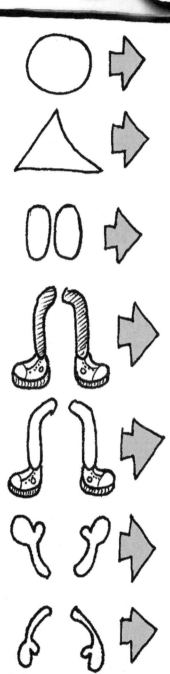

One Hundred Hungry Ants

Dividing 100 hungry ants into different numbers of groups helps students learn the concept of "divide and conquer."

2 groups of 50   5 groups of 20

MATERIALS

- *One Hundred Hungry Ants* by Elinor J. Pinczes
- 5 sets of 100 plastic ants (can be ordered from any novelty company)
- 1 large roll of butcher paper
- Bottle of white glue

PREPARATION

Organize the materials and divide the butcher paper into five large pieces.

STEP BY STEP

Read the book *One Hundred Hungry Ants* to your students. In the beginning of the story, the hungry ants are walking in a straight line, one by one, to the picnic. They realize that they can get to the picnic faster if they divide into two groups of 50, then four groups of 25, then five groups of 20, then ten groups of 10.

Divide your class into five groups and distribute the materials. Have each group represent one of the ways in which the ants divided themselves: two groups of 50, four groups of 25, five groups of 20, etc. Then have each group of students arrange the ants on the piece of butcher paper according to the grouping the students were assigned and glue their ants in place.

In turn, the groups of students explain to the class how they divided to get the end result. The groups then illustrate the picnic, write their section of the story on the butcher paper, and display their work on the classroom wall. Discuss with your students other situations in which it would be helpful to divide. For example, you might want to divide cookies equally among a growing number of guests. (Use the book *The Doorbell Rang* by Pat Hutchins to illustrate this concept.)

VARIATION/EXTENSION

Ask each group of students to write their own version of the story using a different animal or insect. For example, one group might send bears to the beehive to get honey or rabbits to the meadow for clover.

JOURNAL ENTRY

Draw how you would divide 12 cookies onto plates for 4 people.

GRADES: *2–4*   **SKILL ADDRESSED:** *Division*   **VOCABULARY:** *Divide*

ESSENTIAL QUESTION: *How does dividing a large group make a task easier?*

Estimate, Mate!

When students learn to estimate, they develop a valuable life skill.

MATERIALS

- 3 clear, resealable plastic bags in three different sizes
- Large quantities of large and small objects or candy
- 1 small sticky note per child, per week
- Overhead

PREPARATION

This activity is done weekly with the students. Use the smallest bag for the first three months of your estimation process. Resealable plastic bags work well because students can actually manipulate them to estimate the number of objects within. Use large, uniform items with a simple shape to fill the bag the first time, such as large buttons, jellybeans, giant gumballs, etc.

STEP BY STEP

Fill the bag with large items on Monday and display it for the students to view for the week. On Friday, explain that estimating has a lot to do with the skills of observation and multiplication. Place the filled bag on the overhead for 15 seconds. Have the students write down on a sheet of paper the number of objects they think are in the bag. Then ask each one to read her answer and tell how she came up with it.

Place the bag on the overhead again and explain that the best way to estimate is to observe, count, and multiply (OCM). Demonstrate to the students how to observe the size and shape and how to estimate using these observations. Roughly count the number of items in the first row of the bag. Roughly count how many rows there are and multiply the rows by the number of items in the row. Stress to students that the numbers will not be exact. Estimation is a guess.

Come up with a class estimation and write it on the board. Remove the items and count them to determine how close the class estimation was to the actual number.

On the following Monday, fill the bag with different items. Review the estimation process and remind students that they have an entire week to estimate. When ready, they each are to place their estimation and name on a sticky note.

GRADES: 3–6   **SKILLS ADDRESSED:** *Estimation, counting, multiplication*   **VOCABULARY:** *Estimate*

ESSENTIAL QUESTION: *How does estimation help me in everyday life?*

The estimations can be attached to the wall, an estimation bulletin board, etc. On Friday, go over the process once again and monitor success.

As you move to smaller objects and a larger bag, point out to the students that they may not be able to count all of the objects in a particular row, so once again they will have to estimate and multiply.

VARIATION/EXTENSION

As students progress, use nonuniform objects, odd-sized containers, etc.

JOURNAL ENTRY

Describe how you would estimate the number of grandparents who will come to your school's Grandparent Day lunch.

Expanded Notes

Students will understand expanded notation more readily when they can actually expand the note.

MATERIALS

- Sentence strips
- Markers

PREPARATION

Cut each sentence strip in half crosswise, creating two 12-inch lengths. Make enough so that each student has ten 12-inch strips. Create a list of three-place numbers you want to use, writing each either as a numeral or in word form (345 or three hundred and forty-five). Record these on the board.

STEP BY STEP

Distribute the materials and let students work together, if you wish, but explain that each student will be responsible for 10 completed strips. Tell the students to choose 10 numbers from the list on the board and write the numerals in expanded form. For example, if a student chose 345 (or three hundred and forty-five), he would write 300 + 40 + 5 on the strip. Then have them fold the sides of the strip inward (see illustration) to reveal the numeral in its original form (345). Students continue until their 10 strips, each having a different numeral, have been completed.

JOURNAL ENTRY

I can write the 10 numerals on my strips using number words.

GRADES: *2–6*   **SKILL ADDRESSED:** *Expanded notation*   **VOCABULARY:** *Expanded, ones, tens, hundreds*

ESSENTIAL QUESTION: *How can I write a numeral using expanded notation?*

We Are Family

Use this engaging manipulative to teach your students about fact families and watch the fun and learning unfold!

MATERIALS (FOR EACH STUDENT)

- Paper doll patterns (see next page)
- 12 x 18-inch sheet of construction paper
- Markers

PREPARATION

Duplicate the reproducible. Cut each sheet of paper in half crosswise, creating two 9 x 12-inch sheets.

STEP BY STEP

Give each student the reproducible and two halves of the paper. Ask students to fold one into fourths, accordion style. Then ask them to cut out a paper doll shape, trace it onto the folded paper, and cut it out. Be sure to point out, however, that they should not cut around the end of the hands (see illustration: cut only on the solid line). When the paper is unfolded, the students should have four paper dolls that are connected hand-to-hand. Each student then chooses a fact and writes one of the four members of that fact family on each doll. He repeats the activity using the other side of the doll and a different fact family.

Have students repeat the procedure with the remaining half of construction paper, folding it accordion style, tracing and cutting out the other pattern, and writing a different fact family on both sides of the doll.

VARIATIONS/EXTENSIONS

Use two dolls for turn-around facts (6 + 3 and 3 + 6 or 4 + 7 and 7 + 4).

Use two dolls for addition and matching subtraction fact.

Use three dolls for number, set, and number word.

JOURNAL ENTRY

Explain the four parts of a fact family.

GRADES: *1–3*   **SKILL ADDRESSED:** *Fact families*   **VOCABULARY:** *Fact family*

ESSENTIAL QUESTION: *What are the members in a math fact family?*

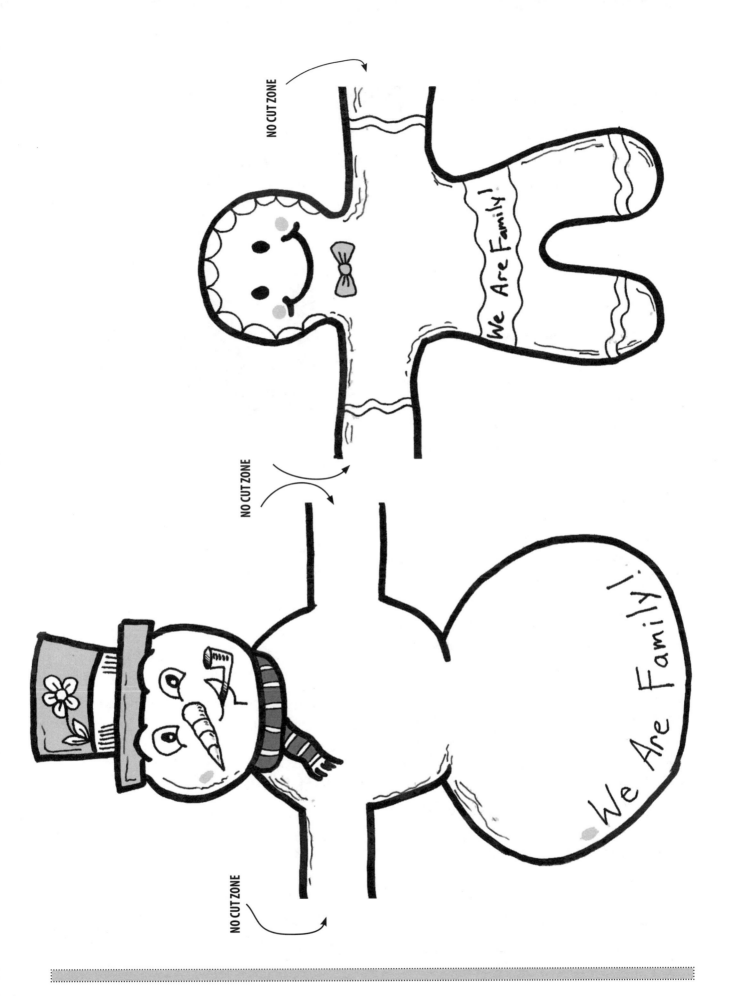

NO CUT ZONE

NO CUT ZONE

NO CUT ZONE

We Are Family!

We Are Family!

Factors and Multiples

This whole-group game enlivens any lesson about multiplication and provides practice with a competitive edge.

MATERIALS

- 2 chairs
- 2 individual dry-erase boards
- 2 dry-erase markers

PREPARATION

Review with your students various multiplication equations and the concept of multiplication as a form of addition.

STEP BY STEP

Write the words *multiple* and *factor* on the board. Explain that a multiple is the result of two numbers being multiplied together. Write the numbers 2 and 3 on the board and ask what the multiple is of these two numbers (6). Repeat with several other sets of numbers until students understand the concept.

Next, write the number 6 on the board and ask what numbers can be multiplied together to equal this number. Students will answer that 3 x 2 = 6 and 6 x 1 = 6. Explain that 3 and 2 are factors of 6, and 6 and 1 are factors of 6.

Divide your class into two teams. Set two chairs in the front of the room and ask one member from each team to sit in one of the chairs. Give both students a dry-erase board and a dry-erase marker. Using the term *multiple*, give the pair a multiplication fact. For example, ask them, "What is the multiple of 10 and 3?" (30). The first team member to write the answer correctly and display it receives a point. Those two students get up and hand their boards and markers to the next two players, who sit in the chairs. Give these two a problem using the term *factor*, such as, "What are two factors of 12?" (6 and 2 or 4 and 3). To keep the game more challenging, don't allow the use of the number 1.

Continue the game until the students are comfortable with the terms and have all had a chance to play.

JOURNAL ENTRY

Write four two-digit multiples and list all the factors of each.

GRADES: *3–6*   **SKILL ADDRESSED:** *Factors and multiples*   **VOCABULARY:** *Factor, multiple*

ESSENTIAL QUESTION: *How are factors and multiples related to multiplication?*

Let's Compare Fractions

Students are able to determine which fraction is greater by watching you work on the overhead. This is especially beneficial to visual learners.

MATERIALS

- Let's Compare Fractions reproducible (see next page)
- Transparency
- Overhead
- Various colors of overhead markers

PREPARATION

Make a transparency of the reproducible.

STEP BY STEP

Place the transparency on the overhead and cover the bottom two fraction circles. Have students count the number of sections in both of the top circles. Point out that each circle has the same number of sections (4). Fill in one of the sections in the first circle with one color of overhead marker. Ask the students to tell you how many sections are filled in (1). Use another marker to fill in two sections of the second circle. Ask students to tell you which circle has more sections filled in and point out that this is the larger fraction. Discuss that the more sections that are filled in, the larger the amount of the circle is filled in, and the larger the fraction is.

Move to the bottom set of circles. Both have six sections. Repeat the process of filling in various sections and discussing which fraction is greater.

VARIATIONS/EXTENSIONS

Invite students to come to the overhead and write the fraction next to the circle.

Introduce the greater than (>) and the less than (<) symbols between the two fractions.

JOURNAL ENTRY

Using the Triman compass, draw two pies and divide each into six equal sections. Fill in a different number of sections in each pie and write the fraction for each.

GRADES: *1–6*   **SKILL ADDRESSED:** *Fractions*   **VOCABULARY:** *Fraction*

ESSENTIAL QUESTION: *How can I tell which fraction is greater?*

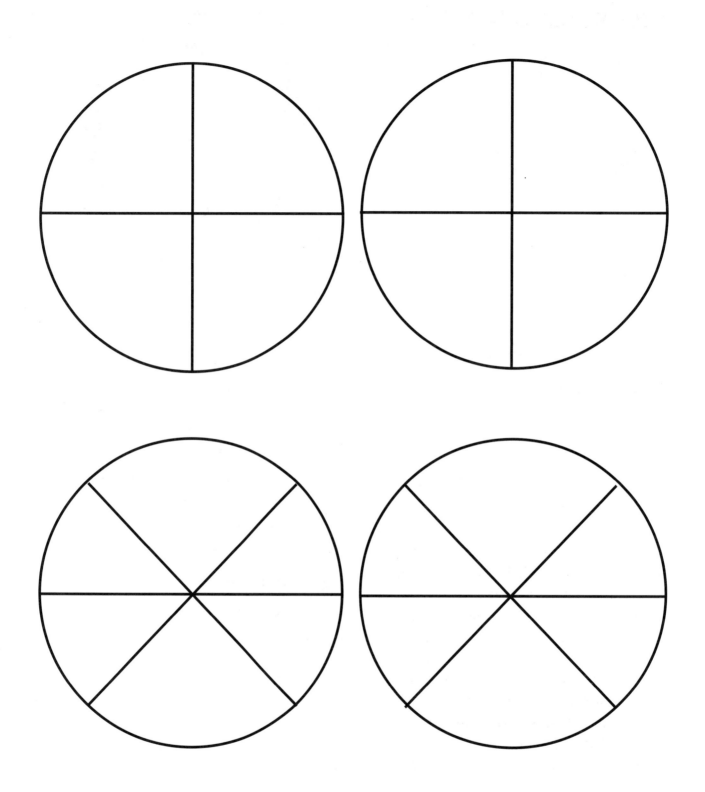

Fractions Are Fun

Ideal for visual and tactile learners, this activity helps kids grasp the concept of fractions and how to reduce them.

MATERIALS (FOR EACH STUDENT)

- Styrofoam egg cartons, many (!)
- Bags of markers (beans, pennies, tiddly winks, small macaroni, etc.)

PREPARATION

Gather your Styrofoam egg cartons. Using scissors, cut off and dispose of the lids. Keep one carton with all 12 cups in tact and cut the others into sets as follows:

2 sets with 2 cups

2 sets with 6 cups

4 sets with 3 cups

3 sets with 4 cups

1 set with 10 cups

2 sets with 5 cups

5 sets with 2 cups

1 set with 8 cups

2 sets with 4 cups

4 sets with 2 cups

12 individual cups

STEP BY STEP

Divide your class into groups of three to four students and give each group a complete set of the prepared egg cartons and a bag of markers.

Ask your students to take the 10-cup egg carton. Write the number 10 on the board under a fraction bar. Explain that none of the egg-carton cups contain a marker, so the fraction of the container that is filled is now 0/10.

Have each group of students add one marker to each cup of their 10-cup egg carton. Ask the students how many cups are filled and write 10/10 on the board. Explain to the students that this 10-cup egg carton is full, and then write the number "1" on the board. Continue to experiment with the 10-cup egg carton by having students put a marker in various numbers of cups and write the resulting fractions on the board with assistance from you.

Practice in a similar way with the 12-cup, 8-cup, 6-cup, etc. cartons. Remind students that the bottom number of the fraction is the total number of cups in the carton and the top is the number of cups that contain a marker. Verbalize as you write the fraction: 2/10 means two-tenths of the carton is filled, 3/6 means three-sixths of the carton is filled, and so forth. Practice this concept daily.

GRADES: *2–4*   SKILL ADDRESSED: *Fractions*   VOCABULARY: *Fraction, group*

ESSENTIAL QUESTION: *How can different fractions be equal?*

Once the students understand the concept of fractions, reinforce it by showing them how to reduce fractions. Beginning with the 10-cup carton, place the two sets of five-cup cartons directly on top of the 10-cup carton. Choose an even number such as 8 and ask students to take 8 markers and see if the markers can be evenly divided between the two sets (4 in each set of 5). Have the students remove one of the five-cup sections and point out that the remaining section is the fraction reduced: 8/10 is reduced and equal to 4/5. Then explain to your students that if the markers cannot be divided evenly, then the fraction cannot be reduced.

VARIATION/EXTENSION

Have students develop their own fraction problems and demonstrate them to the rest of the class.

JOURNAL ENTRY

What would be the best way to divide two candy bars into pieces for three people?

Licorice Stick Geometry

Students learn about shapes, sides, and angles by manipulating licorice sticks. This provides both practice and an opportunity for assessment of a basic unit on shapes.

MATERIALS (FOR EACH STUDENT)

- 10–15 licorice sticks (traditional or strawberry)
- Ruler
- Plastic knife
- Glue
- Card stock

PREPARATION

Assemble all the materials.

STEP BY STEP

Distribute the materials. Explain to your students that this activity will determine if they know the elements of each shape. Tell them that they may use the licorice sticks in any way they want to form each shape. They may cut, bend, or measure them. Ask them first to make a square, using two licorice sticks. Tell students that they will have to cut and measure the licorice so that all four sides are even and the four corner angles are as close to 90 degrees as possible—like the side and bottom of a square. Demonstrate how to do this, and then glue your shape onto the card stock.

After the students have made their squares, have them discuss the specifics of that shape. Give them another shape to create and repeat the process until all shapes have been made and discussed.

VARIATION/EXTENSION

Have students form the shapes without gluing them on the cardstock. Once you have assessed your students, let them eat the licorice. (Check first, of course, with your school nurse about any food allergies or restrictions.)

JOURNAL ENTRY

Make three columns and place the words circle, square, and triangle at the top of each column. Write a list of things that are those shapes.

GRADES: *1–6*   SKILL ADDRESSED: *Geometry*   VOCABULARY: *Shapes you are reinforcing—square, triangle, etc.*

ESSENTIAL QUESTION: *Can I form shapes and learn about sides and angles?*

Geometry Field Trip

After learning about geometric shapes and figures, students go on individual field trips to reinforce their knowledge of the specifics of each.

MATERIALS

- Pictures of flat (one-dimensional) geometric shapes (square, circle, triangle, rectangle, parallelogram, etc.) or wooden models of geometric figures (sphere, cube, cylinder, cone, triangular prism, etc.)
- Various old magazines
- Scissors
- Disposable camera per group of 2 to 3 students, or access to individual digital cameras at the home
- Glue
- Paper for book

PREPARATION

Display pictures of geometric shapes and figures.

STEP BY STEP

Teach lessons on identifying either geometric shapes or figures. Give students opportunities to compare actual geometric shapes or figures with shapes and figures in real life. For example, compare a circle with a round clock face or a cone with an ice cream cone.

Inform students that they are going to create a book for a class of younger (K–2) students. Divide students into groups of two or three and assign randomly a shape or figure (or more than one) to each group. Have the groups go through the magazines, looking for examples of shapes or figures in real pictures and cutting them out for use in the book.

Have students in each group take turns going on geometry "field trips" with parents or older siblings through town, the neighborhood, or even their own homes. Each child should take at least three or four pictures of each shape or figure. Pictures can be printed at home, school, or a photo store.

In class, the groups gather all of their photos and magazine pictures and determine which ones will go on their pages of the book. The pictures are presented to the class and shown how they match the shape or figure. Ask each group to design their page layouts and glue the selected pictures accordingly. Once all the pages from all the groups are assembled as a book, the class presents it to the students in the younger grade.

GRADES: *3–6*   **SKILL ADDRESSED:** *Geometry*

VOCABULARY: *Square, circle, triangle, rectangle, parallelogram, sphere, cube, cylinder, cone, triangular prism*

ESSENTIAL QUESTION: *Can you think of all the geometric shapes in nature?*

VARIATIONS/EXTENSIONS

Have students create posters instead of book pages.

Have students include their own drawings of shapes or figures.

JOURNAL ENTRY

Pick a shape or geometric figure and make a list of all the things you can think of that are that shape.

All Sorts of Things

Here's how to provide hands-on practice with categorizing, graphing, and interpreting data.

Sorting sheet   Graphing sheet

MATERIALS (FOR EACH STUDENT)

- 12 x 18-inch sheet of tagboard
- Items to sort (shapes, buttons, pasta, etc.)
- Graphing sheet reproducible (see next page)

PREPARATION

Make sorting sheets from the tagboard (see illustration) and laminate each one. Gather the items to be sorted and duplicate the graphing sheet.

STEP BY STEP

Distribute the materials. Decide how you want the items to be sorted and then ask students to place the items on their sorting sheet and group them accordingly (shapes by kind, coins by their value, buttons by color, etc.). For example, if students are asked to sort triangles and rectangles, they would place these shapes on their sorting sheets in the appropriate places that you have designated.

Next, have students discuss the information collected and use their graphing sheets to record their data onto their bar graphs. They might note that there are 6 triangles and 9 rectangles, so one bar would be labeled triangles and that bar would be 6 units up on the graph measurement. Another bar would be labeled rectangles and that bar would be 9 units up on the graph measurement. The point of discussion would be how many of each shape do you have, and which is more or less? How many more or less?

VARIATION/EXTENSION

Have students make a picture graph (draw triangles and rectangles, for example) instead of a bar graph.

JOURNAL ENTRY

Summarize today's lesson.

GRADES: *1–6*   **SKILLS ADDRESSED:** *Graphing, categorizing*   **VOCABULARY:** *Bar graph collections, data, sort*

ESSENTIAL QUESTION: *How can I collect data in a way that is clear, concise, and easily understood?*

ALL SORTS of THINGS

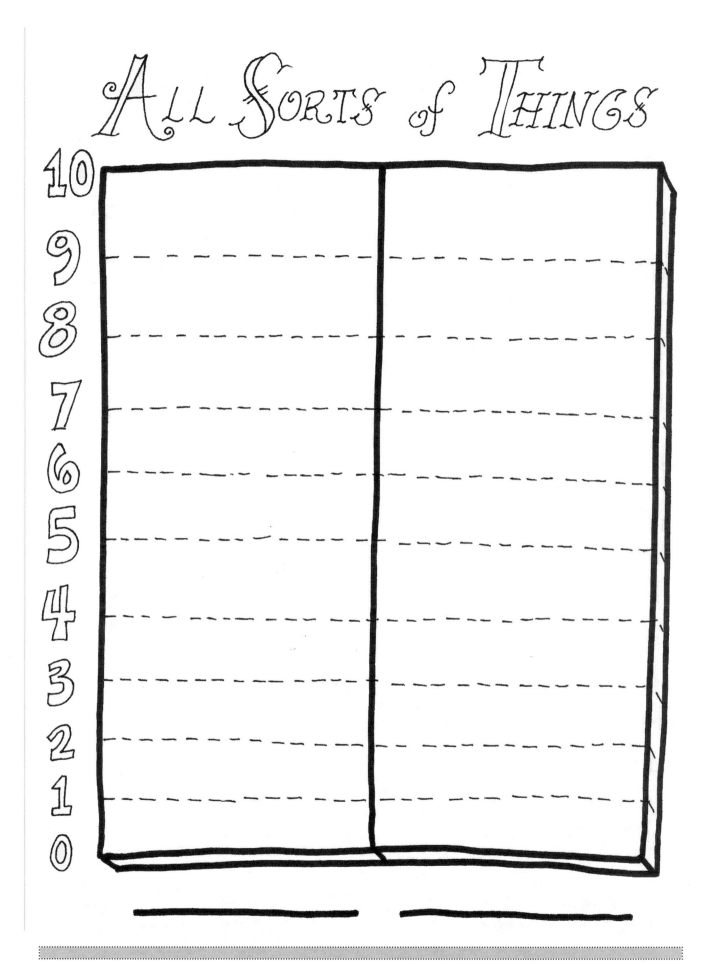

The Talking Graph

Students who can discuss what a graph illustrates truly understand the concept of graphing. This whole-class or small-group activity is extended over two days or two class sessions.

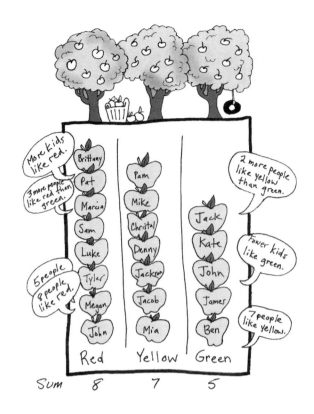

MATERIALS

- Any class-made graph on a 3 x 4-foot sheet of newsprint
- Small index cards
- Markers

PREPARATION

Decide what information will be graphed and what kind of graph will be used (bar, line, circle, or pictograph) and provide the appropriate materials. Cut out a speech bubble from each index card, making enough so that each student has one.

STEP BY STEP

The first day or session, distribute the materials. Have students construct the graph and discuss their findings (specific amounts shown on the graph). Students compare different facets of the graph (which shows more, less, the same). Later, during the second day or session, the graph is revisited. Each student writes in his speech bubble a response to the following: If this graph could talk, what would it say? Students discuss their responses and glue them to the graph in appropriate places according to the information written inside the bubble.

VARIATION/EXTENSION

Have the students work together to create a different kind of graph using the same information.

JOURNAL ENTRY

If you could ask the graph a question, what would you ask?

GRADES: *1–4*   SKILL ADDRESSED: *Graphing*   VOCABULARY: *Speech bubble*

ESSENTIAL QUESTION *What is this graph telling me?*

Let's Make a Deal!

Playing this simple card game, students learn to identify greater than and less than numbers.

MATERIALS

- 1 deck of playing cards for every pair of students

PREPARATION

Remove the face cards and aces from each deck.

STEP BY STEP

Group students into pairs and give each pair a deck of cards. One partner deals the cards, facedown, so that both players have an equal number. Both players place their stacks facedown in front of them. Players take turns turning over the top card on their stacks and the player with the greater number takes both cards. If the numbers are the same, players continue turning cards over until one is greater. Then, the player who has the greater number takes all the overturned cards. Play continues until all cards are gone. The player with the most cards wins. Play the game again but this time, the player with a number less than the other player's takes both cards.

VARIATIONS/EXTENSIONS

Use this same format to learn addition facts. Both players turn over two cards, add them together, and the student with the greater number takes all the cards.

Use the > (greater than) symbol and the < (less than) symbol on index cards and place both cards on the table. Make sure that the cards are positioned correctly so both students are seeing either the less than symbol (<) or the greater than (>) symbol at the same time. Rotate the cards so that students are consistently looking for both greater than and less than.

JOURNAL ENTRY

Tell why you think finding greater and lesser numbers is easier using cards.

GRADES: *K–4*   **SKILL ADDRESSED:** *Greater than and less than*   **VOCABULARY:** *No new words*

ESSENTIAL QUESTION: *Which number is greater?*

How Big Is a Foot?

These practical measuring exercises help students understand the concept of an inch, a foot, and a yard.

MATERIALS

- Ruler for each student
- Yardstick
- 8-inch-long rectangular strip of cardboard for each student
- Jar of quarters
- Jar of pennies
- Jar of dimes
- Copy of *How Big Is a Foot?* by Rolf Myller

PREPARATION

Write the words *inch*, *foot*, and *yard* on the board or on word cards that are posted on the classroom math word wall. Under the word *inch*, place a ruler with one inch highlighted or taped off. Under the word *foot*, place a ruler, and under the word *yard*, place a yardstick.

STEP BY STEP

Distribute the cardboard strips. Tell students to each take out a handful of coins from only one of the jars, and let the children choose the jar they want. Explain to them that they are going to measure their cardboard strips using their coins. Instruct the students to start at one edge of their sheet and lay their coins end to end so that they cover the entire length of the strip. Have each student explain how many coins it took to measure the sheet. Discuss why it took 8 quarters but more pennies and more dimes. Discuss the concept of common measurement. Have students use their rulers to measure the quarter, which is exactly an inch in diameter, and then the cardboard strip. Discuss how having a common measurement will create consistency for everyone.

Read *How Big Is a Foot?* to your students and then discuss how a common measurement tool is important in correctly identifying the size of anything. Have each student measure various objects in the room by inches, feet, and yards using the ruler and the yardstick.

VARIATION/EXTENSION

Measure every child's foot so that students can see how different they are and why a foot must be a common measurement.

JOURNAL ENTRY

How big is a foot? Write how you would measure your own foot.

GRADES: 2–4   **SKILL ADDRESSED:** *Measurement*   **VOCABULARY:** *Inch, foot, yard, ruler*

ESSENTIAL QUESTION: *How can I use any object to really measure?*

Take Away or Tally Up

Using addition is one of the most effective strategies for subtracting mentally, and this game reinforces that connection. This is a great prerequisite to algebra.

MATERIALS

- 2 number cubes (can be made from blank wooden cubes) for each group
- Take Away or Tally Up game board for each player (see next page)

PREPARATION

Make a set of two cubes for each group. Mark one of the following on each face of one cube: 1, 2, 3, 1, 2, and 3. Mark one of the following on each face of the other cube: 4, 5, 6, 7, 8, and 9. Duplicate the game board for each player.

STEP BY STEP

Divide the class into pairs. Give each pair a set of cubes and two game boards. Explain to the students that they are going to use their knowledge of addition to make true subtraction equations. The goal of the game is to complete 12 true number sentences or equations. The first player in the pair rolls the two number cubes and writes the two numbers in one of the number sentences on his game board. The completed number sentence must be true. For example, if Paul rolled a 2 and an 8, he completes the number sentence 10 − 2 = 8 or 10 − 8 = 2. If he cannot make a true number sentence, he makes a tally in the Tally Up box at the bottom of the game board. Then, the other player in the pair repeats the same steps. The first player to complete 12 number sentences before making a total of 10 tallies is the winner.

VARIATIONS/EXTENSIONS

Make a transparency of the game board and play as an entire class. Ask individual students to roll the cubes.

Grades K–1: Use only the numerals 0, 1, 2, and 3 and recreate the game board to have sums of 6 or lower.

Grades 2–4: Mark the following numerals on the second cube: 6, 7, 8, 9, 10, and 11.

After the sentence is completed, write two related addition sentences to go with it. For example, for 10 − 2 = 8, write 2 + 8 = 10 and 8 + 2 = 10.

Have both players share the same board.

JOURNAL ENTRY

Addition and subtraction go together like….

GRADES: *2–6*   **SKILL ADDRESSED:** *Mental math*   **VOCABULARY:** *Addition, subtraction, tally, algebra*

ESSENTIAL QUESTION: *How can I use addition facts to help me subtract?*

Cups o' Beans!

Here's a good way to demonstrate how multiplication is a quick way of counting and adding, which helps students understand this concept.

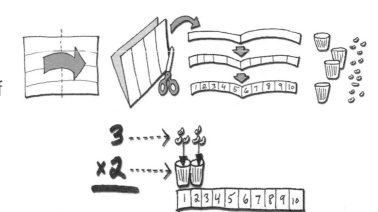

MATERIALS

- Manila or colored folders
- 10 mini cups per student (ketchup cups from fast-food restaurants work great!)
- 1 cup dried beans per student (in resealable plastic bags)
- 1 small dry-erase or chalkboard per student
- 1 dry-erase marker or piece of chalk per student

PREPARATION

Open up a folder and lay it flat. Trim off the tabs, forming a straight edge. Divide the folder horizontally into four equal strips and cut them apart. Divide each strip into 10 equal sections and number them 1 through 10 (see illustration). Make one for each student.

STEP BY STEP

Distribute the materials and ask students to put their numbered strip on their desks. Write the multiplication sentence 2 x 3 = ? on the board. Ask students to place two mini cups on their strip—one in the first section and the other in the second section (see illustration). Discuss the equation and ask how many beans students think should go in the cups. Have each student place three beans in each cup. Point out that the concept is 2 cups times 3 beans. Have the students empty the beans onto both sections. Students will see 3 beans in the first section and 3 beans in the second section. Ask the students to count the total number of beans, which will be 6. Write the total in the equation on the board.

Have students write the multiplication sentence (2 x 3 =) on their dry-erase board or chalkboard and then solve it using the multiplication cups.

VARIATIONS/EXTENSIONS

Reverse the order of the equation (as well as beans and cups) so that students can see that 2 x 3 = 3 x 2.

Varying the objects in the cup (beads, dimes, etc.) gives students a variety of manipulatives to use for repeated practice. Teach the students to count by 2s, 3s, 4s, etc. to learn multiplication tables more quickly. This is a good differentiated activity for those who still need more practice and are above third-grade level.

JOURNAL ENTRY

How is multiplication an easy way to add?

GRADES: *3–5*   **SKILL ADDRESSED:** *Multiplication*   **VOCABULARY:** *Multiply, multiplication*

ESSENTIAL QUESTION: *How does dividing things lead to multiplication?*

Multiplication Mania

This activity reinforces the automatic process and speed of multiplication.

MATERIALS

- Fine-point permanent marker
- 1 small, heavy-duty, resealable plastic bag per student
- 1 multiplication sheet per student (see reproducible on next page)
- 20 pieces of elbow macaroni per student
- One-minute timer for the teacher

PREPARATION

Using a ruler and fine-point permanent marker, draw a vertical line down the center of each plastic bag.

STEP BY STEP

Duplicate the multiplication sheet. Distribute the materials to students. Have each student place six or more pieces of macaroni in his bag and seal it shut. Set the timer to 1 minute and ask students to shake their bags briefly but vigorously. Tell them to lay their bags flat and count the pieces of macaroni on both sides of the vertical line (see illustration). If macaroni pieces extend across the vertical line, students can push them to one side or the other. Using those same two numbers, they write the multiplication sentence on their multiplication

sheet and then write the answer. For example, a student might have placed 6 pieces of macaroni in his bag. After it is shaken and laid flat, the bag might show 2 pieces of macaroni on one side of the line and 4 on the other. The problem to be written will be 2 x 4 = 8. Students then repeat the same process until the minute timer ends. The student with the most correct multiplication sentences is the winner and gets to decide how many macaroni pieces go into the bag during the next round.

VARIATIONS/EXTENSIONS

Let students use dice or spinners to determine the multiplication sentence.

Use a five-minute timer for younger students.

JOURNAL ENTRY

How are multiplication and division related?

GRADES: *2–5* **SKILL ADDRESSED:** *Multiplication* **VOCABULARY:** *Multiplication, multiply, solve*

ESSENTIAL QUESTION: *How can memorizing your times tables help you solve other problems?*

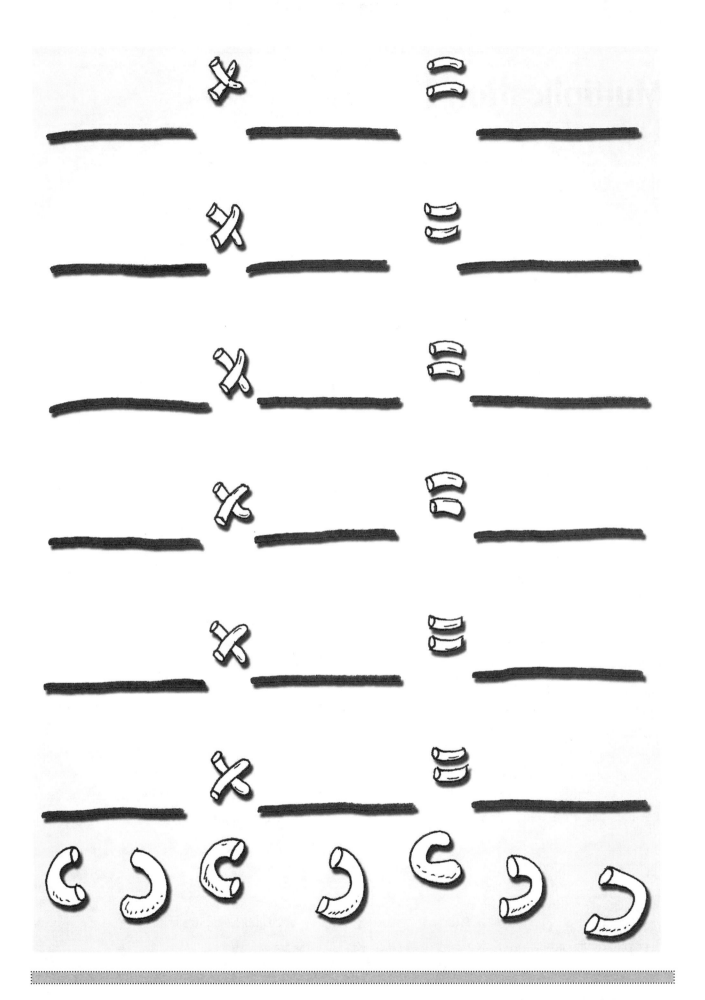

Multiplication Circles

As students repeat and interact with the multiplication tables, they can more easily memorize and learn them.

MATERIALS

- 1 small index card per student or set of multiplication flash cards
- One 3-foot piece of string per student
- 1 small foam-rubber ball

PREPARATION

Either make your own set of multiplication cards with every multiplication fact from 1 through 12 or obtain a set of multiplication flash cards. Whichever you choose, write the answer on the front of each card with a marker. Punch a hole in the top left and right corners of each card and attach both ends of the string through the holes (see illustration).

Decide what multiplication tables you want to work on, such as the 1s and 2s. The day before your lesson, give each student a card with one multiplication fact written on it from the 1s or 2s. Tell the students they are going to become experts on their facts. Explain that they will take their cards home at the end of the day, repeat the fact over and over, and memorize it. The next day, they will bring their cards back to school.

STEP BY STEP

When you are ready to begin the activity, place each student's card around her neck so that every student can view the multiplication fact she is wearing. You should also participate and hang a fact around your neck. Ask half of your students to stand and form a circle, facing the center and standing in random order. Ask the remaining half to sit at the feet of the standing students, also facing the center. You will be part of the standing circle and will start the activity.

While holding the ball, state the fact you are "wearing" and ask the students to repeat it. Then say, "What is my fact?" and have the students repeat the fact. Do this several times for imprinting. Toss the ball to a student and ask her to state her fact. Ask the class to repeat her fact and follow the same line of repetition. As you move from fact to fact, the student tosses the ball to another student so that you can continue the

---

GRADES: *2–4*   SKILL ADDRESSED: *Multiplication*   VOCABULARY: *Multiplication fact*

ESSENTIAL QUESTION: *Will repeating help me remember my multiplication facts?*

line of repetition. Make sure each student in the circle gets to state his fact, and continue to bring in facts already stated.

An example of your dialogue with the circle of students might be as follows: "My fact is 2 x 3 = 6. What is my fact, students? [They respond.] Again. Again. What is my fact? Say it louder. What is my multiplication fact? Mary, what is your fact? Class, what is Mary's fact? Again. Again. What is Mary's fact again? What is my fact? Repeat my fact again. Repeat Mary's fact again. Sammy, what is your fact? Class, repeat Sammy's fact. Julie, repeat Sammy's fact. Class, what is Sammy's fact? Again. Again. Class, what is my fact? What is Mary's fact? Say Mary's fact louder."

Continue with this until each fact in the standing circle has been imprinted. Next, have the standing students change places with those who are sitting and repeat the process. Do this activity at least three times a week using the same tables. The next week move to the 3s and 4s. As your students progress with learning the multiplication tables, mix the tables so that you are constantly reinforcing all of them.

VARIATIONS/EXTENSIONS

Form smaller circles for students who are having difficulty with certain facts and use only those facts on the index cards.

Substitute addition, subtraction, or division facts on the index cards to reinforce these skills.

JOURNAL ENTRY

How does repeating something help you remember it?

The Magic Flasher

Using the "magic flasher" motivates students to learn multiplication and division facts.

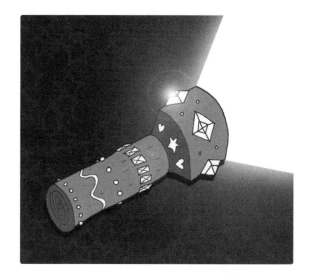

MATERIALS

- Inexpensive flashlight
- Tacky glue
- Glitter
- Plastic jewels
- 18-inch square piece of lamé cloth

PREPARATION

Use glue to decorate the flashlight with glitter, plastic jewels, etc. Place the "magic flasher" in a special location in the classroom and cover it with the cloth.

STEP BY STEP

Develop a series of 10 multiplication or division facts or problems, depending on the level of the skill to be reinforced. Write the answers on the board in random order. Make a production of asking the students if they want to use the "magic flasher."

Slowly uncover the flasher to build excitement. Give the students a problem and tell them the first to solve it will get to use the magic flasher to flash the correct answer. Pass the flasher to that student and ask him to turn it on and flash the answer.

When the problems are all solved, return the magic flasher to its original location and cover with the cloth. To maintain excitement about the "magic flasher," do not use it more than once a week.

VARIATIONS/EXTENSIONS

Affix geometric shapes to the board, read a definition, and have students flash the shape that was defined.

Attach mini classroom clocks to the wall and have students flash the correct digital time from a list of times on the board.

JOURNAL ENTRY

Write a list of math facts you have difficulty with.

GRADES: *1–6*   **SKILLS ADDRESSED:** *Multiplication, division*   **VOCABULARY:** *Focus*

ESSENTIAL QUESTION: *Can using a flashlight to identify a number help us remember?*

Bundles

Bundling drinking straws together provides a visual and tactile way to demonstrate how multiplication and division actually work.

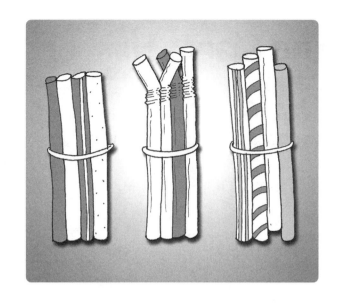

MATERIALS (FOR EACH STUDENT OR PAIR OF STUDENTS)

- 30–50 drinking straws
- 10 small rubber bands

PREPARATION

Decide whether you want students to work independently or in pairs and assemble the materials accordingly.

STEP BY STEP

Distribute the materials. Explain to students that they are going to learn the concept of multiplication and division by bundling straws together. Starting with multiplication, ask students to bundle 4 straws into 3 separate groups and affix each group with a rubber band. Ask the students how many bundles they created with 4 straws in each (3). Then ask them what 4 x 3 equals (12). Have students take the rubber bands off their bundles and count the total number of straws (12), verifying that 4 x 3 = 12. Discuss and practice using various numbers of straws and bundles.

Reverse the process for division by counting out 15 straws and bundling them into groups of 3. Ask students to count the number of bundles that have been created (5). Ask the students what 15 ÷ 3 equals (5). Discuss and practice using various numbers of straws and bundles.

JOURNAL ENTRY

Explain how multiplication is just an easy and quick way to add.

GRADES: *3–6*   **SKILLS ADDRESSED:** *Multiplication, division*   **VOCABULARY:** *Multiply, divide, set*

ESSENTIAL QUESTION: *Is grouping things together a part of multiplying?*

You Choose, I Choose

Dominoes provide practice with the concept of greater than and less than while turning it into a game.

MATERIALS (FOR EACH PAIR OF STUDENTS)

- You Choose, I Choose reproducible (see next page)
- Set of dominoes

PREPARATION

Duplicate the reproducible. Assemble the materials.

STEP BY STEP

Divide students into pairs and distribute the You Choose, I Choose reproducible. Place a set of dominoes facedown between the two players. One student chooses a domino and places it faceup in the space provided on the mat. The student writes the sum of the pips on the line marked 1. Then the other student chooses a domino. If its sum is less than the sum of the first domino, he places it on the jar labeled "Less Than < ." If the sum is greater, he places it on the jar labeled "Greater Than > ." If the sum is the same, he places it on top of the first domino. Players take turns choosing and placing dominoes until all have been used.

The game is played another time, with the other player choosing a domino, placing it in the designated space on the mat, writing the sum on the line marked 2, and so forth.

VARIATION/EXTENSION

Choose two dominoes and add the sums.

JOURNAL ENTRY

Explain what the symbols < and > stand for. Give two examples of each.

GRADES: *1–2*   **SKILLS ADDRESSED:** *Number order, greater than (>) less than (<)*

VOCABULARY: *Greater than, less than*   **ESSENTIAL QUESTION:** *How can I determine if a domino has more or less pips without counting?*

Toss On, Toss Off

Manipulatives and movement provide students with practice calculating addends.

MATERIALS (PER PAIR OF STUDENTS)

- Counters (buttons, pennies, plastic discs, paper clips, etc.)
- 8½ x 11-inch sheet of paper
- On-Off reproducible (see next page)

PREPARATION

Assemble the materials. Depending on which number addend your class is working on, decide how many counters you will need to give each pair of students and write the total on the On-Off reproducible. Duplicate the reproducible.

STEP BY STEP

Divide the students into pairs and distribute the materials. Have the paired students place their sheet of paper on the floor between them. One partner holds the counters in one hand and tosses them into the air over the paper. The other partner records how many counters land on and how many land off the paper. Repeat with partners taking turns tossing. The game is completed when the grid is filled (eight tosses).

VARIATION/EXTENSION

Use higher numbers for a greater challenge.

JOURNAL ENTRY

Write the definition of addend and sum.

GRADES: *1–2*　　　**SKILL ADDRESSED:** *Number sense*　　　**VOCABULARY:** *Addends, sum*

ESSENTIAL QUESTION: *How can I identify the ways to break a total number into two parts?*

GAME 1

Total Number = _____

ON	OFF

TOSS

ON

TOSS

OFF

GAME 2

Total Number = _____

ON	OFF

Happy Trails Hundred Chart

Becoming familiar with the Hundred Chart is a breeze when your students play this guess-my-mystery-number game.

MATERIALS (FOR EACH STUDENT)

- Hundred Chart reproducible (see page 77)
- Plastic disks or markers
- Pencil and paper

PREPARATION

Duplicate the Hundred Chart.

STEP BY STEP

Distribute the materials. Model this game several times before allowing your students to work independently. Choose a particular number on the Hundred Chart but don't identify it. Then tell students to place a marker on a different number that you specify. Have students move their markers according to your directions until they arrive at the mystery number. For example, ask students to place their markers on 25, then move the marker three places to the right, then down one place, and then five places to the left. They will then be at the mystery number, which is 33.

After practicing during several class periods, have the students select their own mystery numbers and then write four clues for

discovering them. Have students exchange clues with each other and find each other's mystery numbers.

VARIATION/EXTENSION

Have students use plus or minus signs with their clues to get to a number on the chart. For example, move to 25 + 3 to the right (28). Or move to 18 + 2 down (38).

JOURNAL ENTRY

Explain how to use the Hundred Chart to find the pattern for odd numbers. Give examples of five odd numbers.

GRADES: *1–4*   **SKILL ADDRESSED:** *Number sense*   **VOCABULARY:** *No new words*

ESSENTIAL QUESTION: *How can I create clues that lead to one particular number on the Hundred Chart?*

Math About Me

This activity makes students more aware of the fact that math is all around us all the time and that so much about us can be mathematical.

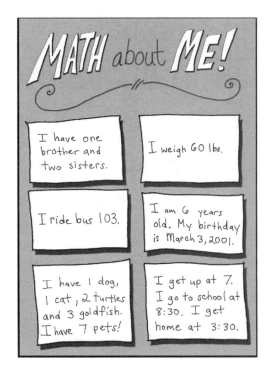

MATERIALS

- *Math Curse* by Jon Scieszka
- 11 x 18-inch sheet of white drawing paper per student
- 8 small, colored index cards per student

PREPARATION

Read the book aloud to your students.

STEP BY STEP

Distribute the materials. Ask students to write at the top of their drawing paper "Math About Me!" Then have them brainstorm examples of personal facts (birth date, height, weight, age, grade, address, number of family members, shoe size, etc.) and daily living facts (wake-up time, portions of food for breakfast, bus number, time spent reading, number of TVs or rooms in the home, etc.) that can be related to math. Students write one or two facts about themselves on each card and then glue the cards on their sheets of paper to create a Math About Me poster (see illustration).

JOURNAL ENTRY

Everything I do is related to math. For example, . . .

GRADES: *1–5*　　　**SKILL ADDRESSED:** *Number sense*　　　**VOCABULARY:** *Height, weight, zip code*

ESSENTIAL QUESTION: *What information about me can be related to math?*

Super-Special Action Subtraction

Combining the fun of competition with the skill of learning subtraction facts equals success for all students.

ACTION SUBTRACTION!

MATERIALS

- Action Subtraction reproducible (see next page) for each pair of students
- Set of dominoes for each pair of students
- Plastic disk or small marker for each student

PREPARATION

Make copies of the reproducible.

STEP BY STEP

Arrange students in pairs and distribute the materials. Tell them to place their set of dominoes facedown. Next, have them put their marker on the "Begin" space on their game board. The first player in each pair selects a domino, determines which of the two faces has the greater value, and subtracts the lower from the greater number. After finding that difference, the player moves her marker that number of spaces toward the "Super Goal." If a player turns over a double, she gets another turn. Then the other player selects a domino, finds the difference between the faces, and moves his marker that number of spaces toward the "Special Goal." The player reaching his goal first is the winner.

VARIATION/EXTENSION

Players can select two dominoes, determine the face value of each, subtract the lower from the greater value, and then move that number of spaces toward the goal.

JOURNAL ENTRY

Write a math word problem. Draw dominoes to help you solve the problem.

GRADES: *K–6*   **SKILLS ADDRESSED:** *Number sense, subtraction*   **VOCABULARY:** *Domino, difference, goal*

ESSENTIAL QUESTION: *Which of the two faces on a domino has the most pips, the least, and what is the difference?*

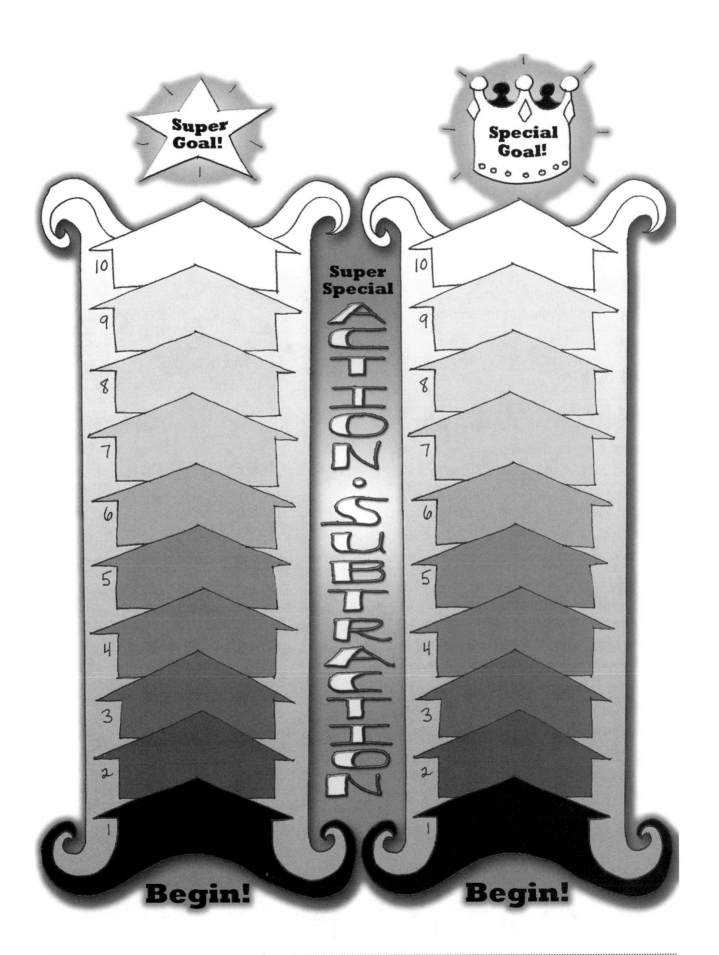

Numbers on the Move

Getting students out of their seats and moving around helps oxygenate their brains, which increases energy and alertness, improves short-term memory, and anchors learning.

MATERIALS

- Sheets of tagboard of different colors
- Markers

PREPARATION

Prepare several sets of numbers for students to order, writing one number on each sheet of tagboard. Make sure you have enough so each student has one. Use a different color for each set. For K–1, prepare multiple sets of the same numbers. For older students, use higher numbers in different sets so that the material is more complex.

STEP BY STEP

Place the numbers in a pile and let each student select one, or just give each child a number. Tell them that when you say, "Merry mix-up," they are to walk freely about the room. Then, when you say, "Numbers on the move," they are to hurry and find others who have the same color sheet as they do and line up in order from least to greatest. See which group can be first to get their numbers in the correct order.

VARIATIONS/EXTENSIONS

Do not use consecutive numbers.

Use numbers in sets of 10, 5, or 2.

JOURNAL ENTRY

Write a counting rhyme, such as "One, two, three, four, five/Angry bees buzz round their hive."

GRADES: *K–4*   **SKILL ADDRESSED:** *Number sense*   **VOCABULARY:** *Greatest, least*

ESSENTIAL QUESTION: *How can I put numbers in order to 100?*

Fill in the Blanks

Students become familiar with the Hundred Chart by filling in the missing numbers.

MATERIALS

- Fill in the Blanks reproducible (see next page) for each student
- White-out liquid or tape
- Dry-erase markers for each student

PREPARATION

Duplicate the reproducible, white out different numbers on each one, and laminate.

STEP BY STEP

Distribute the materials to the students. Ask students to fill in all the missing numbers.

VARIATIONS/EXTENSIONS

Have students fill in five missing numbers, exchange with other students, and fill in five more until the charts are completed.

As students become proficient, use a timer.

JOURNAL ENTRY

Write numerals to 100 by 1s, 2s, 5s, or 10s.

GRADES: *K–6*   SKILL ADDRESSED: *Number sense*   VOCABULARY: *Hundred*

ESSENTIAL QUESTION: *What numbers are missing from the Hundred Chart?*

1	2	3	4	5	6	7	8	9	10
11	12	13	14	15	16	17	18	19	20
21	22	23	24	25	26	27	28	29	30
31	32	33	34	35	36	37	38	39	40
41	42	43	44	45	46	47	48	49	50
51	52	53	54	55	56	57	58	59	60
61	62	63	64	65	66	67	68	69	70
71	72	73	74	75	76	77	78	79	80
81	82	83	84	85	86	87	88	89	90
91	92	93	94	95	96	97	98	99	100

FILL·IN·THE BLANKS!

Let Me See, Which Number Will It Be?

Help your students develop an understanding of numerical order and learn about the process of elimination, which often is used in standardized testing.

MATERIALS (FOR EACH PAIR)

- 24-inch sentence strips
- Sticky notes
- Markers of two different kinds or colors (beans, raisins, buttons, coins, etc.)

PREPARATION

Assemble the materials. Choose a range of numbers according to the level appropriate for your students. For example, use 1 through 10 for kindergartners and 0 through 20 for first through third graders. Each pair will need several markers of two different kinds or colors.

STEP BY STEP

Group students into pairs and distribute the materials. Ask students to create their own number line on a sentence strip using the number range you give them. One partner decides on a secret number from the number line and writes it on a sticky note so he can remember it. He keeps the sticky note out of sight. His partner must try to guess the secret number in as few guesses as possible. After each guess, the partner responds with clues of "too high" or "too low." The player who guesses puts a marker on the number he guessed incorrectly, using one color or kind of marker for the too-high guesses and a different color or kind of marker for the too-low guesses. Using the markers helps narrow the field of possible guesses. When the correct number is discovered, the players switch roles.

VARIATION/EXTENSION

Challenge older students by having them use a number line that begins with 500 or higher.

JOURNAL ENTRY

How did eliminating numbers help you guess the correct number?

GRADES: K–3   **SKILLS ADDRESSED:** Number sense, numerical order   **VOCABULARY:** Elimination, higher, lower

ESSENTIAL QUESTION: How can I use the process of elimination to help me find a missing number?

Hundred Chart Puzzles

These puzzles don't have 100 pieces but they do have 100 numbers! Students can polish their counting skills with this puzzle activity. Use timers and double the fun!

MATERIALS

- Hundred Chart reproducible (see next page)
- Resealable plastic bag for each student
- 1 timer for the teacher

PREPARATION

Duplicate the Hundred Chart, laminate, and cut each apart like a puzzle. Store the pieces for each puzzle inside its own plastic bag.

STEP BY STEP

Present all your students with a bagged puzzle. Challenge them to be the first to reassemble their pieces. Students may also work in pairs. Next time around, set the timer to see how many are able to finish within a certain period of time.

VARIATIONS/EXTENSIONS

For kindergarten students, use only half of the chart (1-50) for the puzzle and cut it into only two or three pieces.

Use this activity for a Math Center.

JOURNAL ENTRY

On the 100th day of school I will bring 100 _____ . I chose these things because _____ .

GRADES: *1–3*   SKILLS ADDRESSED: *Number sense (including numbers 1 to 100), problem-solving*

VOCABULARY: *Hundred*   ESSENTIAL QUESTION: *How fast can I put the numbers in order from 1 to 100?*

Hundred Chart

1	2	3	4	5	6	7	8	9	10
11	12	13	14	15	16	17	18	19	20
21	22	23	24	25	26	27	28	29	30
31	32	33	34	35	36	37	38	39	40
41	42	43	44	45	46	47	48	49	50
51	52	53	54	55	56	57	58	59	60
61	62	63	64	65	66	67	68	69	70
71	72	73	74	75	76	77	78	79	80
81	82	83	84	85	86	87	88	89	90
91	92	93	94	95	96	97	98	99	100

Race to 100

Combine the fun of puzzles, the excitement of a race, and the skill of ordering numbers to 100 with this self-checking partner game!

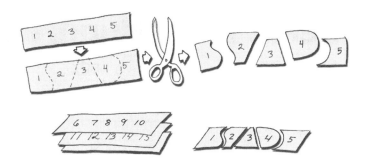

MATERIALS

- Three 12-inch-long sentence strips for each pair of students
- Markers for each pair
- Resealable plastic bag for each pair
- 1 timer for the teacher

PREPARATION

Write a set of five numbers in order on each sentence strip, leaving space between each number. For example, write 1 2 3 4 5 on the first strip, 6 7 8 9 10 on the second strip, and 11 12 13 14 15 on the third. Repeat the process through 96 97 98 99 100. Take the first three sentence strips, making sure they are in numerical order, and cut each strip apart to make puzzle pieces. Place the pieces in a plastic bag. Continue with the remaining sentence strips, again making sure each set of three strips is in numerical order. The number set 1 to 5, 6 to 10, and 11 to 15 should be together in one bag; 16 to 20, 21 to 25, and 26 to 30 in another bag; and so forth through 100.

STEP BY STEP

Group students as pairs and have them work on the floor. Explain that you are going to set a timer and each group will work together putting their three sentence-strip puzzles back together. When time is up, have all the groups join their strips to make one large puzzle from 1 to 100.

Later, challenge your students by giving partners number strips that are not consecutive. For example, one pair might have 1 to 5, 26 to 30, and 86 to 90. Once the partners have assembled their own three strips, they join their strips to the other groups wherever they fit in numerical order.

VARIATION/EXTENSION

For lower grades, have the numbers go only to 50 and allow more time to assemble the pieces.

For upper grades, write numbers in multiples of 5 and use higher numbers from 900 to 1,000.

JOURNAL ENTRY

I can write my numbers to 100 in two different ways.

GRADES: *K–4*   **SKILL ADDRESSED:** *Numerical order*   **VOCABULARY:** *Fifteen, twenty, thirty, forty, hundred*

ESSENTIAL QUESTION: *How can I put numbers in order to 100 using sets of 15 at a time?*

Bears in Line

As students line up their own personal teddy bears, they learn about ordinals and deductive reasoning.

MATERIALS

- Index cards
- 1 teddy bear brought in by each student

PREPARATION

Print a numeral from 1 though 10 and ordinal from first through tenth on an index card. Ask students to line up their bears on a windowsill or other accessible location that you designate.

STEP BY STEP

Choose 10 students and ask them to take their bears from the windowsill. Then have them line up their bears on a table in front of the classroom. Place the index cards with numerals in order on each bear's lap (i.e., the card with 1 on the first bear in line, the card with 2 on the second bear in line, etc.).

Ask the students to count the bears aloud as you or a student holds up the number card. Tell the students that the bears are in line in numbered order, but instead of saying this is bear number one, they should say, "This is the first bear." Replace the "1" card with the "first" card as you do this. Continue down the line so students know there is a different way of identifying each bear's place in the line.

Remove the cards and assess the students by mixing up the cards and having the students (individually or as a group) place the ordinal number card on the correct bear. Repeat the process but use the qualities of a bear to identify the ordinal. For example, you might ask the students to place the correct card on the bear with the blue ribbon and funny smile or on the black bear with the tennis shoes. For reinforcement of the concept, repeat the process daily, using different bears.

VARIATION/EXTENSION

Have students identify a bear by using deductive reasoning. For example, you might say, "This bear has blue eyes." Then the students eliminate the bears without blue eyes by laying them down or turning them so they face backwards, etc. The process continues as you describe different qualities until the correct bear is identified and the ordinal card is placed on the correct bear.

JOURNAL ENTRY

Draw 10 different objects and label each object first through tenth.

GRADES: *K–2*   **SKILL ADDRESSED:** *Ordinals*   **VOCABULARY:** *One, two, three, four, five, six, seven, eight, nine, ten; first, second, third, fourth, fifth, sixth, seventh, eighth, ninth, tenth*

ESSENTIAL QUESTION: *Can I identify where things are in a line?*

Show Me Your Numbers

Here is an easy-to-make game that helps children learn place value, fact families, and simple math facts. It appeals especially to tactile/kinesthetic learners and also works as a quick, no-hassle assessment tool.

MATERIALS (FOR EACH STUDENT)

- Number Cards reproducible (see page 82)
- 1 report card envelope (6 x 9 inches)
- 1 strip of colored tagboard (2 x 9 inches)
- Glue

PREPARATION

Duplicate the Number Cards. Assemble the materials.

STEP BY STEP

Distribute the materials and have students prepare their own envelopes as this is an important lesson in listening and following directions.

Tell students to place their envelopes in front of them, flap side down and positioned horizontally so that the long sides are on the top and bottom. Have them glue their tagboard strips along the bottom edge on the front of their envelopes (see illustration). Point out that they should glue only the bottom edge and the two

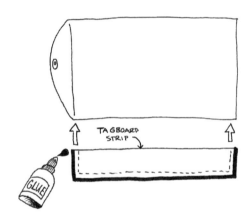

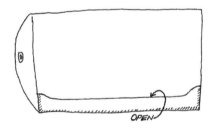

ends of the strip to the envelope so that the top part of the strip is open, forming a long pocket. They do not seal the flap of the envelope.

GRADES: *1–4*   **SKILLS ADDRESSED:** *Place value, fact families*   **VOCABULARY:** *Number, numeral, place value*

ESSENTIAL QUESTION: *How can I show place value for up to four places with a set of numerals 0 to 9?*

Ask students to cut apart the Number Cards and spread the cards out in front of them. Based on grade level, give students a 1-, 2-, 3-, or 4-place number and ask them to construct that number in the pocket of their envelope. Then have them hold their envelope under their chins so that the number is facing them. Wait until all envelopes are under the students' chins and then say, "Show me your numbers." Now students turn their envelopes around so that the number is facing you, and once again hold their envelopes under their chins. Having students hold them this way helps you see when all students are ready as well as make quick observations for assessment. It also keeps cards from flying out of the pockets and getting mixed with other students' cards.

Next, give students two equations in a fact family and have them construct the other two. Review addition, subtraction, and multiplication facts by giving students a problem and asking them to place the answer in the pockets of their envelopes. When everyone is ready, say, "Show me your numbers." Then ask, "What is this number?" Ask which numeral is in the tens place, hundreds place, ones place, thousands place. Once again ask, "What is the number?"

Challenge them further by asking them to change just one of the numerals and make the number 10 less, 100 less, 5 more, 1,000 more, and so forth. Have one student write the number constructed in the pocket on the chalkboard. Students should construct 5 to 10 numbers in one lesson. Finally, look at the numbers listed on the board, read them aloud together, and ask students to arrange them in order from least to greatest.

When finished with the activity, have students store their numbers inside their envelopes.

VARIATION/EXTENSION

Allow students to take turns being the teacher and calling out the numbers.

JOURNAL ENTRY

When I play the "Show Me Your Numbers" game, I feel _____ because _____ .

0	1	2	3
4	5	6	7
8	9	0	1
2	3	4	5
6	7	8	9

The Sock Sale

The four-letter word that can cause a stampede is SALE. Even better are the words BUY ONE GET ONE FREE!! Your students can play this game, shop till they drop, and learn to add and subtract, all at the same time!

MATERIALS

- Masking tape
- Toy or real canned food (6–8 cans per group of students)
- Sale Cards reproducible (see next page)
- 4 or 5 empty, clean tuna fish cans (1 set per group)
- 4 or 5 brightly colored adult socks (1 per tuna fish can)
- Classroom set of plastic coins (enough for 3 or 4 groups of students)

PREPARATION

Write prices, based on the ability of your students, on pieces of masking tape and attach to the tops of the toy or real cans. Duplicate the sale cards, cut them apart, and place 12 inside each tuna can. Then place a tuna can with cards inside each sock. (The can makes the sock sit flat and students can draw cards out of it more easily than out of the sock itself.)

STEP BY STEP

Divide the class into pairs or groups of four or five students and distribute the materials. Tell students that they are going to take turns selecting a canned item to buy. Once each student has decided which can he wants, he reaches into the sale sock and draws a card. That card will tell him the marked-down price. The student then determines the price and counts out enough coins to pay for his food. If the student draws a rebate card, he gets some money back. If he gets the "Buy 1 Get 1 Free" card, he gets another can of food. Players continue until there are no cans left.

VARIATIONS/EXTENSIONS

Kindergarten: Tape a plastic coin to the cans.

Grades 3–4: Use higher prices and sale cards with fractions or percentages (for example, ½ off, ¼ off, 25% off, 50% off).

JOURNAL ENTRY

What could these letters stand for?

S   A   L   E

GRADES: *3–6*   **SKILL ADDRESSED:** *Problem-solving*   **VOCABULARY:** *Sale, buy, purchase, price, rebate*

ESSENTIAL QUESTION: *How can I find the new value of a product after it is on sale?*

2¢ Off	3¢ Off	4¢ Off	5¢ Off
10¢ Off	20¢ Off	25¢ Off	50¢ Off
Buy 1 Get 1 Free	Buy 1 Get 1 Free	Buy 1 Get 1 Free	Rebate 10¢
Rebate 25¢			

Puzzled

Every child likes the challenge of putting a puzzle together. In this quick math-fact review activity, the puzzle is the reward!

MATERIALS (FOR EACH STUDENT)

- 12–15 piece framed puzzle*
- 1 large resealable plastic bag
- Paper and pencil

PREPARATION

Place each unassembled puzzle in a plastic bag. Decide which facts to review.

STEP BY STEP

Have each student number his paper from 1 to 15 or 1 to 12 (depending on the number of puzzle pieces). Call out the review problem and ask him to write the answer on his paper. If he writes the answer correctly, he gets to take one puzzle piece out of his bag and place it in the middle of the puzzle frame. Be sure to point out that students cannot put the puzzle together during the review. Continue calling out review problems until all have been given and all puzzle pieces are out of the bags. Now give students 5 minutes to put the puzzles together. Since they have not seen the puzzle assembled, the challenge is on!

VARIATION/EXTENSION

Have students work as a team to solve more difficult problems or multi-step problems and then work together to put a 50-piece puzzle or large floor puzzle together.

JOURNAL ENTRY

Write a word problem using one of the review facts.

These puzzles have a frame around the area where the pieces are put together.

GRADES: *K–6*   SKILL ADDRESSED: *Problem-solving*   VOCABULARY: *Solve*

ESSENTIAL QUESTION: *How can I use a puzzle to help me with math facts?*

Problem of the Day

Your students can become proficient and independent problem solvers with a quick problem-of-the-day activity.

MATERIALS

- 3 sheets of blank paper per student
- Stapler

PREPARATION

Prepare a step book of your own first using the directions below, so when you model this for your students, you will already understand the process.

STEP BY STEP

Distribute the sheets of paper. Modeling as you go, give students the following directions: Stack the three sheets of paper together so there is a front, middle, and back sheet. Move the front sheet <u>down</u> so that its top edge is 1½ inches below the top edges of the other two stacked sheets. Then, keeping it in place, move the back sheet <u>up</u> so that its top edge is 1½ inches above the middle sheet of paper (see illustration).

Keeping all three sheets in their positions, fold them together crosswise, creating a book with a total of six steps. Staple the pages together in three places along the folded edge.

Once the step books have been created, ask your students to write their names and the current date on the top step. On the remaining five steps (the overlapping flap), have them write the days of the week in descending order, from Monday through Friday.

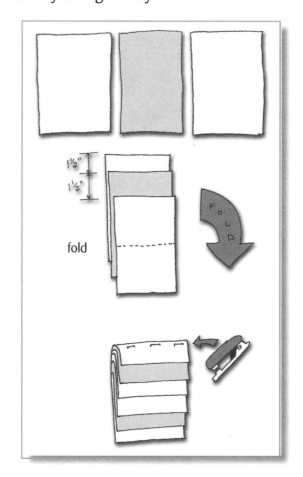

GRADES: *1–6*   **SKILL ADDRESSED:** *Problem-solving*   **VOCABULARY:** *Problem, solve, step book*

ESSENTIAL QUESTION: *Can you solve one question each day this week all by yourself?*

Ask students to lift the top step (or flap) of their books and write, solve, and illustrate the problem of that day. When they have finished, have students share their strategies with classmates. Then go over the strategies and answers with your class.

Examples of Problem of the Day:

✎ Continue these patterns: (show different shapes in a repeating pattern)

✎ Continue these patterns: 5 10 15 20

  — — — — — —

✎ Write the fact family for 12.

✎ Draw a geometric robot. Include one square, four circles, three triangles, and eight rectangles.

✎ How many combinations of coins can you make that will equal 25 cents?

✎ Write the numbers 1 through 50 and circle the even numbers.

✎ Fill in the blanks with one of the numbers from 0 to 12 so that the total of each row is the same. Use 6 two times. Don't forget turn around facts (9 + 3, 3 + 9).

___ + ___ = 12   ___ + ___ = 12   ___ + ___ = 12

___ + ___ = 12   ___ + ___ = 12   ___ + ___ = 12

___ + ___ = 12   ___ + ___ = 12   ___ + ___ = 12

___ + ___ = 12   ___ + ___ = 12

✎ Starting at 5:30 PM, if you spent two hours doing your homework and 45 minutes eating dinner, what time would it then be? Draw the three clock faces showing the time you began, the time you completed your homework, and the time you finished eating.

✎ Place one addition symbol and one subtraction symbol between the following numbers to make an equation that equals 12.

6   2   8 = 12

VARIATION/EXTENSION

In addition to using your own word problems, allow students to create their own.

JOURNAL ENTRY

Make up a problem with 4 numbers, so that the first 3 numbers equal the 4th number, based on where you insert an addition and a subtraction symbol between the 3 numbers.

Hundred Chart Magic

Students use the Hundred Chart for creating equations and problem-solving as they discover the magic sign.

MATERIALS (FOR EACH STUDENT)

- 5–10 plastic chips or markers
- Hundred Chart reproducible (see page 77)
- Pencil and paper

PREPARATION

Duplicate the reproducible and assemble the materials.

STEP BY STEP

Distribute the materials. Ask students to place 5 to 10 chips on their Hundred Chart in a formation of their choice, which will be their magic sign. For example, it might be an X shape, square shape, triangle shape, M shape, + sign, etc. Then, as the students remove each chip from its place, they write a clue to its position—one clue for each chip. Have them exchange their charts with another student who will use the clues to discover the magic sign (see illustration).

VARIATION/EXTENSION

Increase the number of chips and clues for older students.

JOURNAL ENTRY

Write about an activity that could be done on the one hundredth day of school.

GRADES: *1–6*   SKILL ADDRESSED: *Problem-Solving*   VOCABULARY: *No new words*

ESSENTIAL QUESTION: *What clues can I write that will help others find my magic sign?*

Skittles Ratios

Working with Skittles, students can visually determine ratios.

MATERIALS (PER STUDENT OR PAIR OF STUDENTS)

- Index cards
- Small bag of Skittles
- Dry-erase markers

PREPARATION

On the blank side of the index card, make a ratio mark in the center and then laminate each.

STEP BY STEP

Distribute the materials. Have each student place the ratio card on the center of his desk. Then have him divide the Skittles into one pile of each color—purple, green, yellow, orange, and red. Direct the students to place 2 yellow pieces on the left side of the ratio mark and 3 green pieces on the right side. Explain that the ratio of yellow pieces to green pieces is 2 to 3 or 2:3. Now, ask students to create an equal ratio by doubling the pieces of both colors of candy—4 yellow: 6 green. Have students create other ratios using different colors of candy. Then ask them to make equal ratios by doubling, tripling, or quadrupling the number of Skittles.

VARIATIONS/EXTENSIONS

After forming the ratios, have the student write the ratio on his card.

Encourage students to create word problems using the ratio card and Skittles.

JOURNAL ENTRY

How would you use ratios in your everyday life?

GRADES: *1–6*   **SKILLS ADDRESSED:** *Ratios, fractions*   **VOCABULARY:** *Ratio, equal, double, triple, quadruple*

ESSENTIAL QUESTION: *How can using Skittles as manipulatives help me learn about ratios?*

Cube It!

Using dice and a skills cube, students learn that practice makes perfect.

MATERIALS (PER STUDENT OR PAIR OF STUDENTS)

- Tagboard
- Cube pattern reproducible (see next page)
- Glue
- Set of dice
- Dry-erase board
- Dry-erase marker

PREPARATION

Trace around the pattern on a piece of tagboard, making enough for each student or each pair of students.

STEP BY STEP

Distribute the materials. Have each student or pair cut out the cube pattern from the tagboard. Decide what skills you want to reinforce and ask the students to label each side of their cube with one of the operation symbols. For example, if you want to reinforce addition, subtraction, multiplication, and division, students write each operation symbol on one side of the cube, repeating them until the cube is full. Then they assemble their cube, securing the sides with glue (see illustration).

Tell students to roll their cubes to determine which skill will be reinforced. Then have them roll the set of dice for the numbers to be used. For example, if multiplication is rolled on the cube and 6 and 3 are rolled on the dice, the child will solve the problem 6 x 3 on the dry-erase board. Walk around the room to check for accuracy.

Allow older students to roll the dice twice to process double-digit numbers. For example, if a student rolled multiplication and then 4 and 3 and 1 and 6, the problem to be solved would be 43 x 16, 34 x 61, 43 x 61, or 34 x 16.

VARIATION/EXTENSION

Have older students use x^2, x^3, square root, etc. on the cube.

JOURNAL ENTRY

Which math skill do you think you should practice most? Why?

GRADES: *1–6*   **SKILL ADDRESSED:** *Reinforcement of math skills*   **VOCABULARY:** *Digit and double-digit*

ESSENTIAL QUESTION: *Will practicing help me learn?*

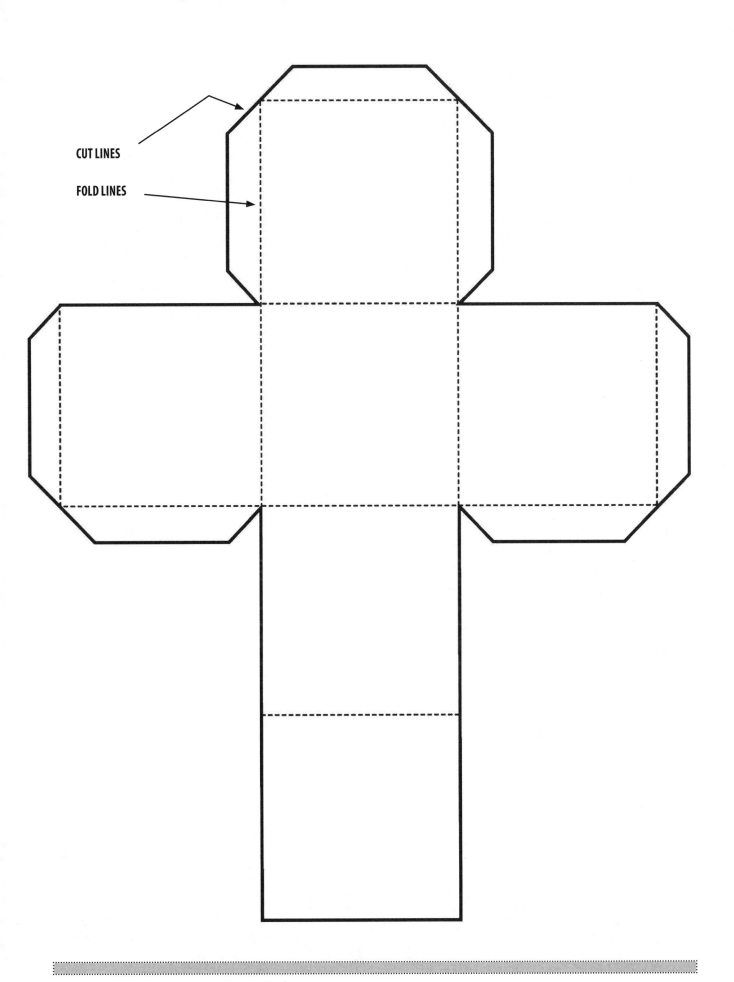

CUT LINES

FOLD LINES

Math Scavenger Hunt

Increase your students' awareness of real-world mathematics with this team math scavenger hunt.

MATERIALS (FOR EACH STUDENT)

- Magazines
- Newspapers
- Math Scavenger Hunt reproducible (see next page)

PREPARATION

Duplicate the reproducible.

STEP BY STEP

Divide students into teams of three or four and distribute the materials. Have each team read the Math Scavenger Hunt sheet, decide how they are going to divide the responsibilities, and figure out where to look for each item on the list. Depending on resources and time, this may be a two-day assignment so that students may also look at home. Students get one point for every item found.

VARIATION/EXTENSION

Have students identify the math concept with each of their findings and explain how it is used in real life.

JOURNAL ENTRY

The three hardest things to find on the list were _____ .

I finally found them _____ .

GRADES: *1–6*   SKILL ADDRESSED: *Reinforcement of math skills*

VOCABULARY: *Statistics, advertisement, pictograph*

ESSENTIAL QUESTION: *Where can our team find the things listed on our Math Scavenger Hunt sheet?*

1. A picture of a fraction
2. A sale with 20% off
3. A centimeter ruler
4. A picture of a car selling for more than $15,000.00
5. A rectangular prism
6. A cylinder
7. A cone
8. A pyramid
9. A line graph
10. A bar graph
11. A pie graph
12. A food label with nutrients listed
13. A pattern in reading material, glued onto paper, and extended
14. An advertisement that includes measurement
15. A math-related children's book
16. A postage stamp
17. A parallelogram
18. A rhombus
19. A trapezoid
20. Parallel lines
21. Perpendicular lines
22. Lottery odds
23. A map
24. A list of statistics for a sports team
25. Something dated 10 or more years ago
26. A sphere
27. Foreign currency or coin
28. An item weighing less than 1 ounce
29. An item weighing more than 5 pounds
30. An item other than a ruler that can be used to find measurements

Extra, Extra! Read All About It!

Children in the upper elementary grades will enjoy using the newspaper for a math scavenger hunt.

MATERIALS (PER STUDENT OR PAIR OF STUDENTS)

- Newspaper
- Marker

PREPARATION

Gather newspapers, create a list of things to be found (or use the checklist below), and write the list on an overhead or on the chalkboard.

STEP BY STEP

Distribute the materials. Explain to students that you are going to show them a list of certain things you want them to look for in the newspaper. When they find one, they circle it and look for another.

List of things to look for:

- » percent sign
- » weather symbol
- » number between 50 and 100
- » sports statistic
- » symbol or word for inches, feet, or yards
- » graph
- » number less than 10
- » schedule
- » something that comes in 2s, 3s, or 4s
- » number greater than 100
- » triangle
- » days of the week
- » year
- » number that is more than 100 but less than 1,000
- » price
- » item that is on sale
- » circle
- » fraction
- » time
- » picture of a geometric solid shape (cube, cylinder, sphere, etc.)

Give students 30 minutes to complete the activity. The one who finds the most items is the winner.

JOURNAL ENTRY

The newspaper is full of math….

GRADE LEVEL: *4–6*   **SKILL ADDRESSED:** *Reinforcement of math skills*

VOCABULARY: *Symbol, schedule, statistics*

ESSENTIAL QUESTION: *How many things pertaining to math can I find in the newspaper?*

Have a Ball

Turn this symbol for fun into a tool for math practice, using a little imagination and a marker!

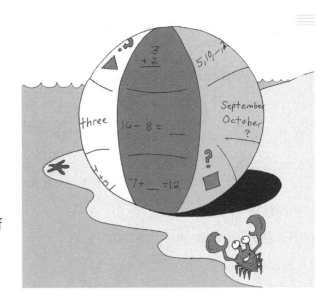

MATERIALS

- 16-inch striped beach ball for each group of four students
- Fine-point permanent marker

PREPARATION

Divide the striped areas of the beach ball into thirds, using the marker (see illustration). You can create an all-subjects ball by writing varied types of questions within each section: math facts, shapes, number words, math vocabulary, clock faces, coin combinations, greater than and less than equations, pre-algebra equations, geometric shapes, and any other math concept you would like to practice or review. Or, you can designate one ball for each type of practice, such as one with addition facts, another with subtraction facts, another with multiplication facts, etc.

STEP BY STEP

Divide the class into groups of four. Explain that each player will gently toss the ball to one of her group members. The student who catches the ball answers the equation where his right thumb is on the ball. Continue the game for about 10 minutes. Move among the groups, observing and making anecdotal or mental notes on students' progress.

VARIATION/EXTENSION

Send a ball home for parents to practice math facts with their children. This will make homework painless for the parent, profitable for the student, and fun for all.

JOURNAL ENTRY

How do you feel about this activity?

GRADES: *K–3*   SKILL ADDRESSED: *Reinforcement of math skills*   VOCABULARY: *No new words*

ESSENTIAL QUESTION: *How can I use a beach ball to help me practice math facts?*

Aim for the Stars

Students practice subtraction facts with a partner in this dice game.

MATERIALS (FOR EACH PAIR OF STUDENTS)

- Star reproducible (see next page)
- 2 dice
- Plastic chips or markers

PREPARATION

Duplicate the Star reproducible, making enlarged copies for younger students.

STEP BY STEP

Divide the students into pairs. Provide each pair with a Star sheet, two dice, and plastic chips or markers. The students take turns rolling a die and subtracting the number from 10. Students locate that same number (the difference) on their Star sheets and cover it with a chip or marker. Play continues until all stars are covered.

VARIATION/EXTENSION

Make the numbers on the stars higher and use two dice for grade 2.

JOURNAL ENTRY

Choose three subtraction facts for 10. Write a related addition fact for each one.

GRADES: *K–2*   **SKILL ADDRESSED:** *Subtraction*   **VOCABULARY:** *Difference*

ESSENTIAL QUESTION: *How can I use subtraction to determine what number combinations make 10?*

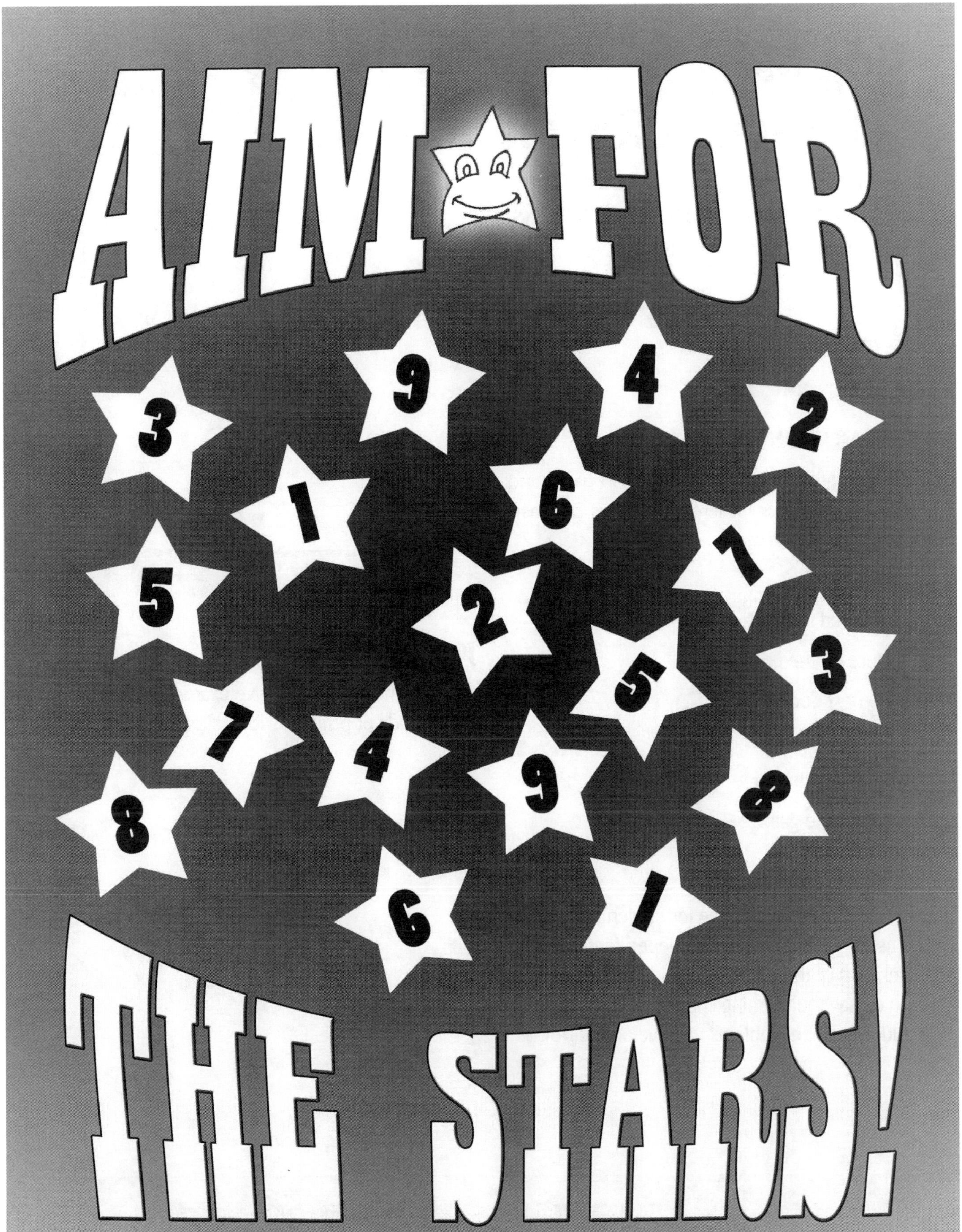
AIM FOR THE STARS!
3
9
4
2
1
6
5
7
2
5
3
7
4
9
8
8
6
1

Sing Me a Subtraction Rhyme

Renaming and regrouping can be simple for your students when they learn this easy-to-chant rhyme of rules.

MATERIALS

- Chart paper

PREPARATION

Write the following rhyme on chart paper and post it in your room where all students can easily see it and refer to it.

More on top?

No need to stop!

More on the floor?

Go next door.

Get ten more!

Numbers the same?

Zero is the game!

STEP BY STEP

This is a great resource for students to refer to as they solve problems independently. Model each part of the poem with a subtraction problem as you say the rhyming lines out loud. Then give students other problems to solve on their own.

JOURNAL ENTRY

Write the rhyme and give examples of the three kinds of problems identified.

GRADES: 2–6   SKILL ADDRESSED: Subtraction   VOCABULARY: Rename, regroup, zero

ESSENTIAL QUESTION: How can I be sure when to rename and when to regroup when solving a subtraction problem?

The Human Clock

A human clock helps students understand the concept of telling time to the hour, half-hour, quarter-hour, and five minutes.

MATERIALS

- Card stock of different colors
- Yellow and blue markers
- Hole punch
- 24 lengths of yarn or ribbon, all about the same length
- 2 lengths of yarn or ribbon, one shorter than the other

PREPARATION

On 12 different cards, all of the same color, write each hour of the day—1:00, 2:00, 3:00, 4:00, and so forth through 12:00. On 11 cards of another color, write five-minute intervals—:05, :10, :15, :20, :25, through :55. On the quarter-hour cards (:15 and :45 cards only) draw a large yellow dot, and on the half-hour card (:30 card only) draw a large blue dot. Punch a hole in the top left and right corners of each card. Tie one end of a piece of yarn to one corner and the other end to the other corner. Cut out two arrow shapes, both of a different color, and attach one arrow to one end of the long piece of yarn and the other arrow to the shorter piece.

STEP BY STEP

Ask 12 students to stand in the shape of a clock. Hang one of the hour cards around the front of each student's neck and one of the five-minute cards around the back of 11 students. Pick another student to stand in the middle of the circle (or clock), holding the short piece of yarn with an arrow on the end (the hour hand) in one hand, and the longer piece of yarn with arrow (the minute hand) in the other hand. Ask the remaining students to sit at the feet of the students forming the clock. Explain that you are going to demonstrate the concept of telling time by using the students as the actual hours and minutes of the clock.

Write a time on the board, such as 3:00. Ask a student sitting on the floor to help you set the yarn hands to this time: the long hand (the

GRADES: *2–5*   SKILL ADDRESSED: *Time*   VOCABULARY: *Hour, minute, time*

ESSENTIAL QUESTION: *How can becoming part of the face of the clock help you tell time?*

minute hand) points to the student wearing the 12:00 card and the short hand (the hour hand) points to the student wearing the 3:00 card. This shows the time to be 3:00.

Write the time 4:45, for example, on the board and ask another student to help set the clock to the quarter-hour. The student sets the hour hand so that it points to the student wearing the 4:00 card. Now ask the other "clock" students to turn around so the student setting the time can see the five-minute, quarter-hour, and half-hour cards. The student then points the minute hand toward the student wearing the :45 card.

Next, do five-minute intervals using the same procedure. This really helps students understand the concept of time.

The human clock should be used consistently, practicing all elements of time: hour, half-hour, quarter-hour, and five-minute intervals. The various colors of the signs and the colored dots help students distinguish the minutes from the hours.

VARIATIONS/EXTENSIONS

Vary which students are part of the clock, set the time, and tell the time.

Give students small individual clocks to use to set the time.

JOURNAL ENTRY

How does it help you to know what time it is?

Time-Telling Teams

The best way to learn to tell time is to practice being the clock!

MATERIALS

- 12 x 18-inch sheets of tagboard
- Clock face reproducibles (see pages 102–103)
- Brads
- Hole punch
- 24-inch pieces of yarn
- Dry-erase markers and erasers

PREPARATION

Make clock bibs for half of your class, using the tagboard and reproducibles. Laminate. Attach the hands with brads. Punch a hole in the top left and right of the clocks as indicated. Attach both ends of the yarn through the holes. Use a dry-erase marker and eraser for the digital watch.

STEP BY STEP

Divide the class into pairs. Have one partner wear the clock bib and the other set the time. Partners stand facing each other. Ask students to set the clocks to a certain time depending on grade level and ability. For kindergartners and first graders, begin with the hour only. Second and third graders may begin with elapsed-time settings of a quarter hour. The partner will set the hands on the clock accordingly. The student whose partner is wearing the wristwatch will write in the digital time with a marker. After several examples, the students exchange places and the process is repeated.

VARIATIONS/EXTENSIONS

For kindergartners, staple the minute hand to the 12 so that the students manipulate only the hour hand.

Use elapsed-time problems for second graders and above. An example might be, if you begin your homework at 6:00 and work for 1½ hours, what time would your homework be completed?

JOURNAL ENTRY

(Use a clock stamp) Write four sentences, one for each time setting. Tell what you might be doing at that time.

GRADES: *K–3*   **SKILL ADDRESSED:** *Time*   **VOCABULARY:** *Hour, half-hour, quarter-hour*

ESSENTIAL QUESTION: *Can you set a clock to tell the time to the hour, half-hour, and quarter hour?*

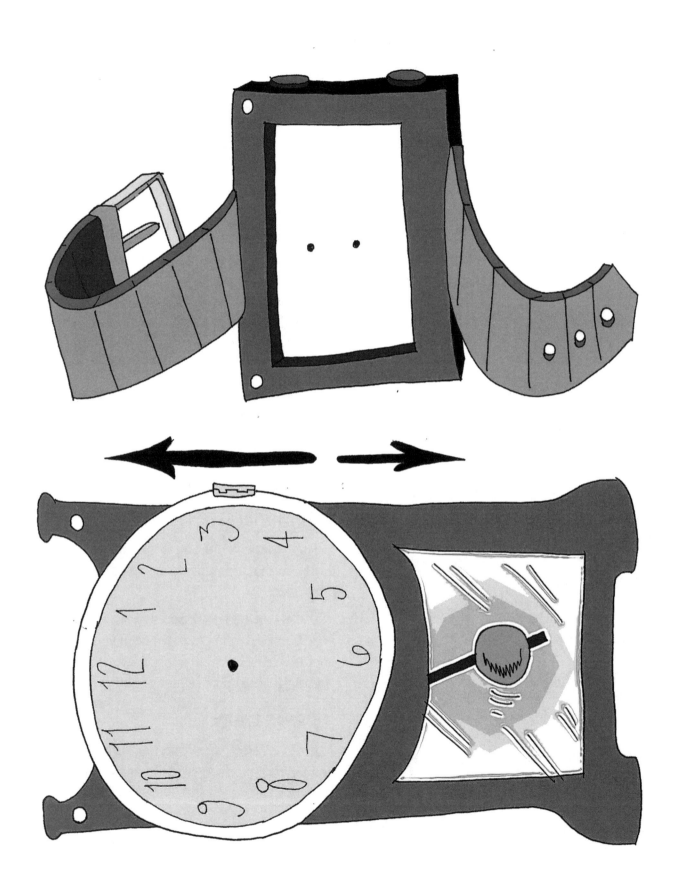

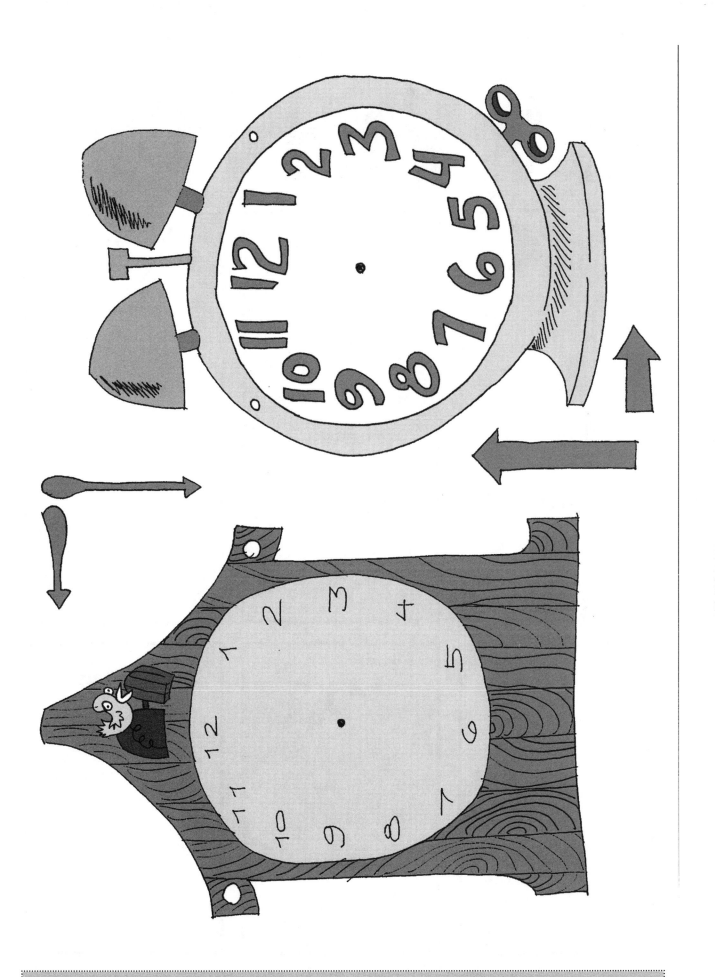

Math Word Folder

A a

B b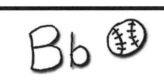

C c

D d

E e

F f

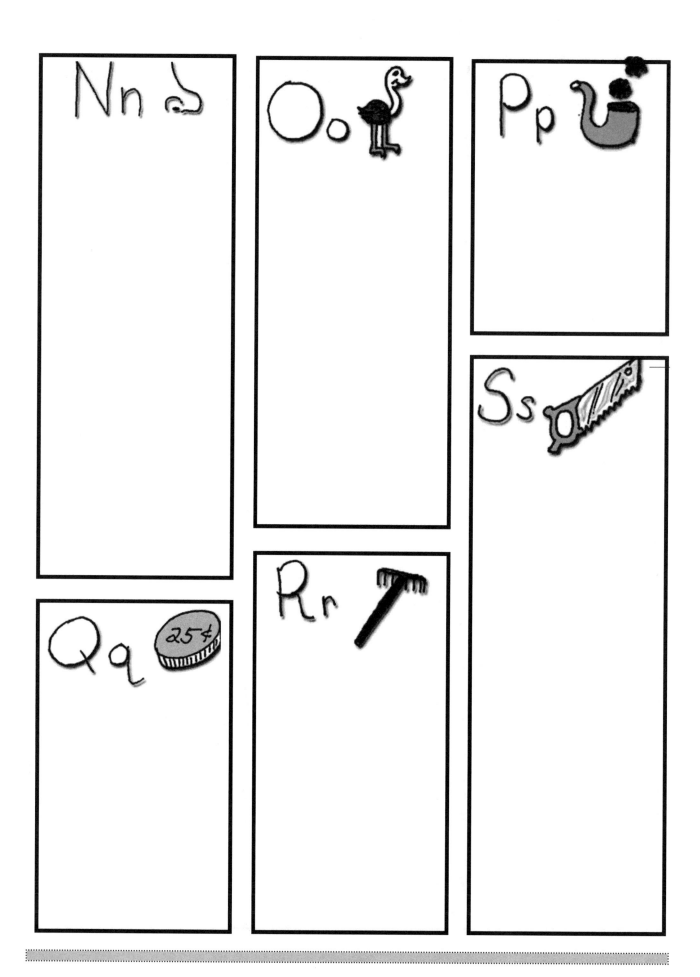

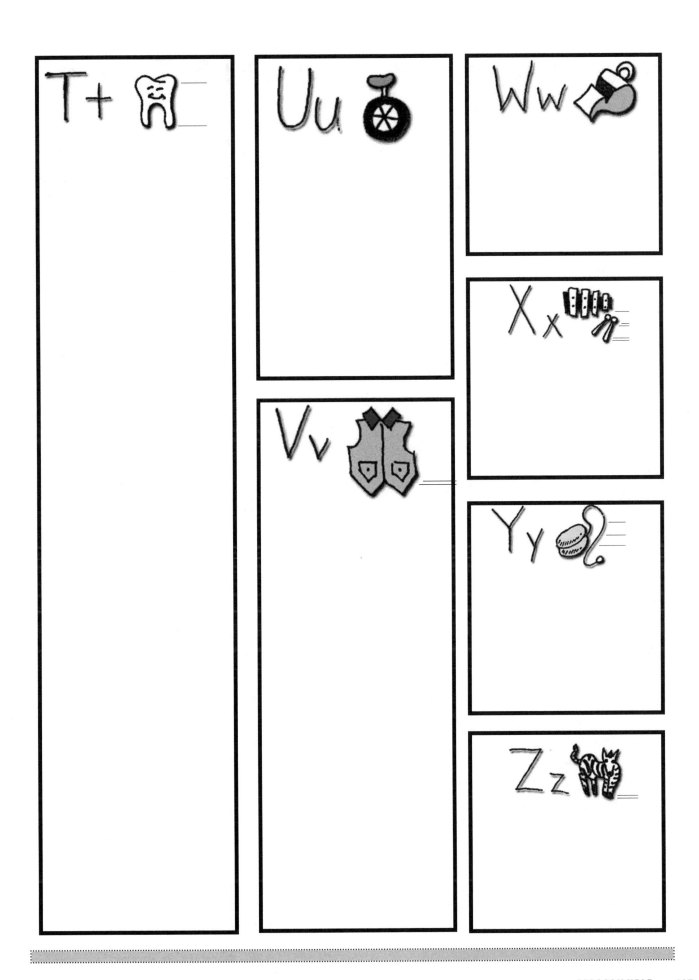

Recommended Reading (Math Books)

Title	Author	Skill
Ten in a Bed	M. Rees	Beginning addition/subtraction
Each Orange Had 8 Slices	P. Giganti, Jr.	Counting
Arctic Fives Arrive	E. Pinczes	Counting
100th Day Worries	M. Cuyler	Counting
Five Little Chicks	N. Tafuri	Counting
Anno's Counting Book	M. Anno	Counting
The M&M Counting Book	B. McGrath	Counting/graphing
A Chair for My Mother	V. Williams	Decimals/estimating/money
A Remainder of One	E. Pinczes	Division
Betcha!	S. Murphy	Estimating
How Hungry Are You?	D. Napoli/R. Tchen	Fractions
The Hershey's Milk Chocolate Fractions Book	J. Pallotta	Fractions
Fraction Action	L. Leedy	Fractions
Apple Fractions	J. Pallotta	Fractions
Polar Bear Math	A. Nagda	Fractions
Fraction Fun	D. Adler	Fractions
Gator Pie	L. Mathews	Fractions
The Doorbell Rang	P. Hutchins	Fractions
Inchworm and a Half	E. Pinczes	Fractions
Grandfather Tang's Story	A. Tompert	Geometry
Color Zoo	L. Ehlert	Geometry
Shapes	J. Reiss	Geometry
Sir Cumference and the Dragon of Pi	C. Neuschwander	Geometry
Sir Cumference and the Sword in the Cone	C. Neuschwander	Geometry
Sir Cumference and the Great Knight of Angleland	C. Neuschwander	Geometry
Mummy Math	C. Neuschwander	Geometry
The Greedy Triangle	M. Burns	Geometry
What's Your Angle, Pythagoras?	J. Ellis	Geometry
When a Line Bends…A Shape Begins	R. Greene	Geometry
Tiger Math	A. Nagda	Graphing
A Million Dots	A. Clements	Large numbers

Mathematickles!	B. Franco	Math concepts
How Big Is a Foot?	R. Myller	Measurement
Inch by Inch	L. Lionni	Measurement
Millions to Measure	D. Schwartz	Measurement
Super Sand Castle Saturday	S. Murphy	Measurement
The Rajah's Rice	D. Barry	Measurement/multiplication
Arthur's Funny Money	L. Hoban	Money
Henry's Pennies	L. McNamara	Money
Dollar and Cents for Harriet	B. Maestro	Money
How Much Is a Million?	D. Schwartz	Number concepts
If You Made a Million	D.Schwartz	Number concepts
Pigs at Odds	A. Alexrod	Probability
Probably Pistachio	S. Murphy	Probability
The Most Wonderful Egg in the World	H. Heine	Shapes
A Nice Walk in the Jungle	N. Bodsworth	Shapes
Sam Johnson and the Blue Ribbon Quilt	L. Ernst	Shapes
Ten Black Dots	D. Crews	Shapes
The Village of the Round andSquare Houses	A. Grifalconi	Shapes
My Full Moon Is Square	E. Pinczes	Solving equations
Math Curse	J. Scieszka	Solving equations
The Jolly Postman	J. & A. Ahlberg	Spatial relationships
The Paper Crane	M. Bang	Spatial relationships
Rooster's Off to See the World	E. Carle	Spatial relationships
Rosie's Walk	P. Hutchins	Spatial relationships
Wing-A-Ding	L. Hoopes	Spatial relationships
Big Time Bears	S. Krensky	Time
Five Minutes' Peace	J. Murphy	Time
The Glorious Flight	A. & M. Provensen	Time
The Grouchy Ladybug	E. Carle	Time
Sunday Morning	J. Viorst	Time
The Sun's Day	M. Gerstein	Time
The Very Hungry Caterpillar	E. Carle	Time

Index

Note: *Page numbers in italics refer to reproducibles to be used with activities.*